THE NEUROSCIENCE
BEHIND SCIENCE FICTION

AUSTIN LIM, PhD

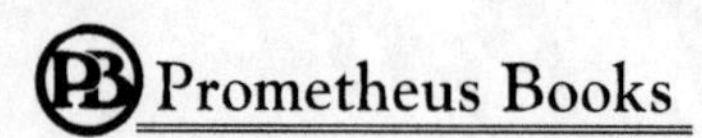

An imprint of The Globe Pequot Publishing Group, Inc.
64 South Main Street
Essex, CT 06426
www.globepequot.com

Distributed by NATIONAL BOOK NETWORK

British Library Cataloguing in Publication Information available

Library of Congress Cataloging-in-Publication Data available

ISBN 978-1-4930-8478-4 (paperback)
ISBN 978-1-4930-8479-1 (ebook)

™ The paper used in this publication meets the minimum requirements of American National Standard for Information Sciences—Permanence of Paper for Printed Library Materials, ANSI/NISO Z39.48-1992.

CONTENTS

INTRODUCTION

> Learn from me, if not by my precepts, at least by my example, how dangerous is the acquirement of knowledge, and how much happier that man is who believes his native town to be the world, than he who aspires to become greater than his nature will allow.
> —Mary Shelley, *Frankenstein*

There's a reason that Mary Shelley's *Frankenstein* (1818) is a staple of literature and pop culture. It's a groundbreaking novel that blended gothic horror, dark fantasy, and otherworldly science gone awry, thus defining early science fiction and laying the literary tracks for future works of horror. It's packed full of evergreen themes such as revenge, nature versus nurture, the meaning of humanity, and the search for love and companionship, all concepts that have resonated with audiences since its publication. It is the inspiration for many works over the centuries across different media, including radio dramas, movies, ballets, video game characters, and even a Victorian burlesque play. Plus, it's an excellent reminder to stay humble: Frankenstein was written by a teenager on summer vacation.

Beyond the narrative of the story, *Frankenstein* is about curiosity. Curiosity is a powerful force that motivates science, so it is not surprising that Shelley's embodiment of the spirit of wonder, protagonist Dr. Victor Frankenstein, is an inquisitive person whose key reason for being is the discovery of truth about nature. Both real and fictional scientists are driven to understand the mysterious phenomena they observe in the universe. They are constantly seeking more knowledge about the world around them, and through use of trial and error, they arrive at answers, which, more often than not, opens the door to further questions. But curiosity is not just a trait relegated to scientists. It is a basic human feature. There are far more scientists in the world than just those who do science for a living.

Curiosity exists only because of the unknown. Conveniently, the unknown is also the birthplace of horror. Even before literacy, people had been making haunting "what-if" stories about the biggest unknowable: what happens after death. The ancient Mesopotamians wondered what if a rogue spirit from the underworld escaped or a person did not receive a proper burial. According to their traditions, those unlucky souls become ghosts, trapped on Earth to torment the living with sickness and misfortune. Some traditions ask what if our good deeds are tabulated and used to determine the next body that your eternal soul will inhabit. If your accumulated karma is bad, then your new life will face all the suffering brought on by your previous life. There's a fear in knowing that your bad deeds and thoughts will follow you beyond your physical body or that you are the reason that crappy things keep happening.

Considering all these questions about the afterlife that permeate cultures around the globe, it's no wonder that Dr. Frankenstein is most motivated to conquer death, like many real-life research scientists of today. Chemists and transplant surgeons have developed medications and procedures that extend life or restore the fullness of life to those for whom illness or injury took a part of theirs. Geneticists have found tools that allow for the control of

DNA and mRNA, offering the ability to shape generations of mice in our image, manipulating physical traits such as fur color and behaviors (e.g., how often they clean their faces). Evolutionary biologists have discovered that we shared common ancestors with other animals, questioning the doctrine that souls are given exclusively to humans.

The questions explored by today's scientists are much more granular and incremental than the "what-ifs" hinted at in speculative fiction, such as reanimation and faster-than-light travel. But there is still an unsettling realization that our curiosity into the secrets of the unknown might reveal some disturbing truths about the world. Curiosity might not directly kill the cat, but it might kill the younger brother, best friend, and wife-to-be.

Space: the final frontier. —Captain Kirk, *Star Trek: The Original Series*

Both horror and science fiction, whether concerned with life beyond death or our tiny place in the universe, evoke a sense of awe. The brilliant technological ingenuity of *Homo sapiens*, with our meager attempts to stave off death, is diminished into nothingness when looking at the timescale and expanse of the universe, billions of years old and billions of light years across. Space is full of mysteries and unknown horrors, sure to both fire up the imagination and instill wonder.

But instead of looking upward at the vast openness of space above, look within for a moment. You'll find something even more amazing inside your head. It's a three-pound lump of fatty jelly.

The microscopic-level features of the adult brain put up some impressive, awe-inspiring numbers. Modern neuroanatomists estimate that the brain contains more than 86 billion (86,000,000,000, or 8.6×10^9 in scientific notation) neurons. These cells are specialized for communicating, using electrical signals called action potentials and chemical signals called neurotransmitters. Neurons exist in different shapes and sizes, each one specialized for an appropriate function. Neurons form complex connections with other neurons called synapses; a single neuron might connect with more than a few hundred thousand other neurons. Altogether, you have an estimated 125 trillion (125,000,000,000,000, or 1.25×10^{14}) synapses in your brain alone. Amazingly, the brain gets all these connections to form in the right places guided by a probabilistic process of cellular trial and error. Most of this precision wiring happens in utero and during the first few years of life.

To add an even deeper layer of complexity, these neurons and the synapses between them are not static once they are laid down. Instead, they are ever changing in subtle but tangible ways. Throughout our lives, we fine-tune these trillions of connections, strengthening some while weakening others. This capacity for change is called plasticity. Suppose, as most neuroscientists do, that the brain uses plasticity as the substrate for learning and that the neural changes take place at the level of some molecular machinery. In theory, these tiny changes might be thought of as a single unit of binary information—a bit, using terminology from the world of computing. Estimates suggest that the typical synapse can exist only in enough meaningfully different states such that each synapse can store 4.7 bits of memory. So we can estimate how many "units" of data our brains can store by multiplying 4.7 by the estimated number of synapses in the brain, putting our memory at around a couple petabytes, or a few quadrillion (2,000,000,000,000,000, or 2×10^{15}) bytes of memory—not bad considering that the price of a terabyte hard drive with 1 trillion bytes (1×10^{12}) is a little less than $100 as of the time of writing.

More generous estimates imply that the brain is less like a classical digital device and more like a quantum supercomputer, where each synapse can store values between 0 and 1. In this more powerful model of neurons, the storage of data happens not only at the level of the synapse but also at the level of individual molecules around the synapse, from the receptors that mediate intracellular communication to the proteins that serve as the skeleton that maintains the complex shape of the cells. If true, the amount of data storage in the brain is a wildly incomprehensible figure, somewhere in the hundred billinovemoctingentillion range, or $10^{8,432}$ bytes. This number vastly overwhelms the number of stars in the universe, at a paltry 3×10^{23}, or even the number of atoms in the universe, at approximately 10^{86}.

To say that the human brain is complex would be the understatement of a lifetime. Our human experience is tiny not only in relation to the universe that surrounds us but also when we try to dive into the limitless depths of our own minds. Unsurprisingly, neuroscience is one of the most puzzling of the modern fields of study as scientists attempt to understand the near-infinite expanses contained within the real final frontier, the one inside your head. Modern molecular neuroscience has only confirmed what Emily Dickinson already knew in the 1860s. The brain is much, much, much wider than the sky.

After all the gaps in our present knowledge have been filled in, imagine the scary "what-if" that lies at the end of all the questions. What if our entire human experience is a consequence of our brains? This leads to the disconcerting thought that we are at the mercy of a bunch of chemicals released by cells that we have little control over. Creativity, the great lengths people go to in search of love, those random moments when we suddenly become aware of our own mortality, the fear of missing out (FOMO), that tiniest irrational inkling of a thought that maybe one of your friends actually doesn't like you, and, of course, curiosity are such nuanced and complex facets of our life. These, some of our most human characteristics, are encoded by a bunch of neurons ticking out a binary signal of ones and zeros.

But it is the next logical step that takes us to a dreadful place of utter helplessness. People have gambled themselves into overwhelming debt, risking their family's financial well-being on a literal die roll, despite intellectually knowing that the house always wins. People have dosed themselves with cartons worth of toxic cigarette smoke for decades, fully aware of the risks. People have carried out mass killings in the name of carefully engineered concepts, such as nation or God. At any moment, there could be maybe a neuron or maybe a billion neurons already in place that could prevent these people from doing these things. Nevertheless, the person stays their destructive course. The haunting realization at the end of neuroscience is that we are not even in control over our own bodies and that our fears, desires, emotions, consciousness, and everything else in the multifaceted human experience are guided by clumps of cells.

> The most merciful thing in the world, I think, is the inability of the human mind to correlate all its contents. We live on a placid island of ignorance in the midst of black seas of infinity, and it was not meant that we should voyage far. The sciences, each straining in its own direction, have hitherto harmed us little; but some day the piecing together of dissociated knowledge will open up such terrifying vistas of reality, and of our frightful position therein, that we shall either go mad from the revelation or flee from the light into the peace and safety of a new dark age.
> —H. P. Lovecraft, *The Call of Cthulhu*

This book is a tour of the brain and its myriad features in all its complexity—the good and the bad, the normal and the atypical—guided by a selection of horror and science fiction works throughout literature, movies, television, and pop culture. It's an exploration of the science that inspired or can be found among the fiction. It's a peek into the mad scientists who perpetrate bizarre crimes and the fears that haunt doomed protagonists. It's a detective story that diagnoses characters with rare psychiatric conditions. But I'm not claiming to try to explain all the events of fiction through neuroscience. I can't demystify the supernatural, nor would I want to. Instead, I only wish to share the knowledge I have collected as a forever student of neuroscience and psychology, using the stories and characters from the genre's best-known works as guideposts.

Within the following pages are tiny bits of insight into the human mind. And while there is much more happening in our brains than we will know in our lifetimes, each piece brings us ever closer to comprehending our existence and all the baggage that comes along with that. I am hoping to fan the flames of curiosity about brain science in the same way that Shelley's *Modern Prometheus* does, even if it means revealing some dark shadows along the way.

NOTES ABOUT WRITING

There is a frustrating amount of tentative language ahead, lots of might's and possibly's. Biology rarely deals in absolutes, so when you spot one of those elusive never's in the book, just notice that it probably sits adjacent to specific caveats, such as "thus far published in the scientific literature." Relatedly, there is also a necessary but regrettable amount of simplification throughout these pages. Writing about science leaves me feeling as if I am balancing on a knife's edge. One terrible outcome is that by omitting important details of an experiment, I misrepresent the life's work of a fellow scientist and distort the truth about nature. The other equally undesirable outcome is that I bore you. Any work of writing geared toward audiences outside of an extremely niche specialization requires this fine line to be carefully traversed. But if you are deeply interested in specifics, maybe the reason why a measurement of hemodynamic response through functional magnetic resonance imaging (fMRI) is not a true assay of how neuronal firing changes, why it is so difficult to use genetically modified mice to learn about the human condition, or any other topic brought forward through these chapters, just know that there are excellent reviews on PubMed written by experts in their respective fields who have spent their entire research careers exploring these nuances.

Generally, everything in neurobiology is more detailed than I have space to explain or, more honestly, time to learn about. I struggle to think of any examples in the brain where the relationship is as simple as "the function of brain area X is to control behavior Y." Any given X does lots of things, and many interconnected networks of brain parts are responsible for any given Y. Further complicating the story is that new techniques for studying the brain are being developed all the time, allowing for more granular subdivisions for each X, meaning that structures previously just called X are more accurately described as an X1, X2, X3, and so on. Additionally, there's that pesky "chicken or egg" situation to contend with. Sure, a bigger X might lead to more Y, but there's also the possibility that repeatedly doing Y leads to a bigger X. All this is to say, be skeptical. Always. It makes you a better scientist and an educated citizen, two things we desperately need in the world.

Regarding footnotes, those attached to literary or movie references are likely to contain spoilers. I'll stay vague when I can, but sometimes the big ending or the dramatic twist is the most essential piece of the story I'm trying to tell. Most of the endings are long past their "spoiler statute of limitations" anyway. You probably already know Lieutenant Ripley defeats the Queen at the end of 1986's *Aliens*, and you've had more than a century to read *The Turn of the Screw*. Rest assured that even if I reveal the fate of the characters, the journeys are just as thrilling as the destination.

CHAPTER ONE

THE OLDEST AND STRONGEST EMOTION*

NEURAL SUBSTRATES OF FEAR

HORROR AUTHORS ARE A VARIED BUNCH. THEIR THEMES RUN THE gamut of ideas, from traditional gothic depictions of how past traumas manifest as the unexplainable to the speculative science fiction imaginings of artificial intelligence (AI) gone rogue. Even with such a diversity in topics, all the greatest storytellers share one thing in common: all are masters of psychology. They manipulate readers with words, transporting us into claustrophobic crypts and abandoned spacecrafts. They introduce us to colorful characters, then make us suspicious of their long-forgotten family secrets and mysterious hidden agendas. And when they successfully leverage the evocative power of fear, their creations stick in our heads long after we are done reading.

In many ways, instilling a sense of dread using the written word is a huge challenge. A clump of bulbous eyes and writhing masses of slimy tentacles is only as weird as the reader's imagination. Authors must put a lot of faith in their audience, trusting that they can turn adjectives into imagery. If the reader doesn't picture themselves as a prisoner in an Eastern European castle far from home, a small-town

* Part of a quotation by H. P. Lovecraft, a leader of the weird fiction movement. The rest of the quotation reminds us that fear of the unknown is the greatest fear, a theme that recurs through much of his storytelling.

resident participating in a strange annual ritual, or a visitor to a decaying East Coast town full of vaguely odd-looking people, then the sense of tension never builds, and the big reveal loses its punch.

But in another sense, fear is easy. Of the complex web of emotions that motivate animals to do things, fear is the most primitive and most thoroughly ingrained in our being. Fear served the most successful organisms since ancient evolutionary times, driving behaviors that promoted their survival (and therefore reproductive) likelihood. Without fear, some early critter might have swum straight into the jaws of a toothy-mouthed hunter. Imagine its kin, a different critter that knew to stay still beneath the ancient seaweed while the predators passed by. For the next millions of years throughout the evolution of species, a healthy sense of fear has increased the odds survival.

> Dishonored and tragic, she was all before me; but even as I fixed and, for memory, secured it, the awful image passed away. Dark as midnight in her black dress, her haggard beauty and her unutterable woe, she had looked at me long enough to appear to say that her right to sit at my table was as good as mine to sit at hers. While these instants lasted, indeed, I had the extraordinary chill of feeling that it was I who was the intruder. It was as a wild protest against it that, actually addressing her—"You terrible, miserable woman!"—I heard myself break into a sound that, by the open door, rang through the long passage and the empty house. She looked at me as if she heard me, but I had recovered myself and cleared the air. There was nothing in the room the next minute but the sunshine and a sense that I must stay.
> —Henry James, *The Turn of the Screw**

* *The Turn of the Screw* (1898) describes the plight of a governess, two children, and two ghosts. James paints an unreliable narrator and dances with ambiguity as to whether the ghosts are real. Stephen King wrote that *The Turn of the Screw* is one of the two greatest supernatural novels ever written. The themes and plot have been reimagined in different media throughout the years, most recently in Mike Flanagan's *The Haunting of Bly Manor* (2020).

Fear is a normal, healthy response that activates robust arousal mechanisms throughout your whole body. These mechanisms work at two levels, the physiological and the cognitive. An influential theory of fear processing, formally described in 1996, calls the physiological response the "low road" and the cognitive element the "high road." According to this framework, when we experience something scary, signals get sent along both pathways. The quicker pathway is the low road, which is the primitive survival response that protects the body from potential harm. This is automatic, so we have no conscious control over these circuits and how they influence the body. Then the second high road pathway fires up, which is the more complex, "hold on a second, let's think this through" response. This is when the cognitive, rational processing dampens low road activity.

Consider this framework in light of Henry James's *The Turn of the Screw*. Seeing the ghost of the late governess haunting the mansion where you work and screaming in response: low road. Automatic. Scream to scare it away. Live to be a Victorian governess another day. Thinking that dead people don't materialize out of thin air or that it is now your duty to protect the innocent children placed at your charge: high road. Complex. Of course ghosts don't exist. This is just a psychotic hallucination of latent sexual repression.

Physiologically, the fear response increases blood flow to our muscles, speeds up our heart rate, and quickens our breathing. We may get goosebumps, possibly a holdover from our much hairier evolutionary ancestors, causing us to look bigger or more intimidating. Our body also reacts in ways that we aren't usually sensitive to, such as the slowing of our digestive processes, the ramping up of certain enzymes, and the dilation of our pupils. These involuntary physiological changes are due to the sympathetic nervous system, a series of nerves that increase the concentration of the signaling neurohormone norepinephrine in the bloodstream and all over different organs. Norepinephrine prepares our bodies for quickly moving to action. Do we fight, throwing a punch at the pale white figure who inexplicably materializes in front of us? Or do we turn, running out of the haunted old mansion, never looking back? Whichever course of action we decide on, the thumping in our chest and the pulsing in our blood vessels help us succeed.

The sympathetic nervous system gets its cues from a brain structure called the hypothalamus. This area sends signals down through the spinal cord, activating nerves that trigger the sympathetic nervous system response. The hypothalamus also functions in the closely related experience of stress, which is a sort of "diet fear," activating many of the same physiological changes but to a lesser degree and on a timescale of weeks and months rather than seconds. There are other features of the hypothalamus of some interest, including the regulation of internal homeostatic states, hunger, and sex drive. As the joke goes, there are four Fs of the hypothalamus: fight, flight, feed, and reproduction.

Tracing the pathway of fear farther upstream of the hypothalamus is the amygdala,* a pair of almond-shaped structures buried deep in the brain within the temporal lobe. And, in turn, the amygdala receives signals from many, many brain regions, such as the following:

- The pulvinar, a subdivision of the thalamus, integrates visual information with whole-body responses. Of particular interest to the 51 percent of you ophidiophobes out there, the cells of the pulvinar seem to be particularly sensitive to snakes. Signaling from the visual centers of the brain through the thalamus and then to the amygdala is

* Many brain structures are paired, and the amygdala is no exception. There is a left and a right amygdala, one in each half of our brain. The grammatically correct plural is "amygdalae," but the convention in the field has been largely to refer to paired structures using the singular form.

a relatively quick circuit, so the low road is fired up at the sight of serpents, harmless or not.

- The prefrontal cortex (PFC), the fanciest part of our brain, is important for long-term planning and helps us do the more difficult thing when faced with a set of choices. Circuits here help us inhibit our base animal impulses. This is the theoretical origin of the high road.
- The orbitofrontal cortex, the clumps of brain tissue just above the eyeballs, gets strong inputs from the visual system pathways, which helps us figure out whether we can believe our eyes. Many times, the answer is no (more about this in chapter 8).
- The inferotemporal cortex is heavily implicated in visual memory associations and object recognition. Is that the figure of a person standing in the window of a faraway tower? Maybe. Or it could just as well be someone's laundry, hung up to dry, billowing in the breeze.
- The insula, sometimes called the "fifth lobe" of the brain, cares a great deal about disgust in both the physical sense ("Yuck, a person covered from head to toe in rashes and fresh scabs") and the moral sense ("Yuck, a woman who seduced her own father to have sex with him").*

All these connections come with numerous complex caveats. It should begin to feel that the brain is a tangled but extremely precise crisscrossing of cables. It's confusing enough to drive someone mad.

* In a double whammy of insula activation, according to poet Dante Alighieri in his *Divine Comedy*, the former was the fate of Myrrha of Greek mythology for committing the latter. In *Inferno*, the most famous section of his epic trilogy, Dante is given a guided tour of Hell and bears witness to the various tortures that befall the souls of the damned. He also puts himself beside Virgil, his literary idol, and describes the violent torment of people who pissed him off in real life, making it technically a piece of fan fiction.

No! the commotion began again. There was a rustling and shaking: surely more than any rat could cause. I can figure to myself something of the Professor's bewilderment and horror, for I have in a dream thirty years back seen the same thing happen; but the reader will hardly, perhaps, imagine how dreadful it was to him to see a figure suddenly sit up in what he had known was an empty bed. —M. R. James, *Oh, Whistle, and I'll Come to You, My Lad**

Given that so many pathways involved with fright intersect or signal through the amygdala, it has naturally become the central hub of interest for people who study the neurobiology of fear. Part of this focus has to do with the case study of Patient SM, a person for whom the neural circuitry of fear doesn't work the way it's supposed to. A survivor of a rare degenerative disorder called Urbach-Wiethe disease, she began experiencing a calcification of her medial temporal lobe early in childhood. As the disease progressed, the accumulation of excess biological waste caused focused damage, specifically destroying both amygdalae of her brain while leaving the rest of the brain mostly intact. Neuropsychologists found in SM a truly unique loss-of-function case study, allowing them to explore the question "How would a person without functioning amygdala behave?"

The first studies of SM tested her ability to recognize fear in others by looking at their faces. A dozen similar-aged participants served as the comparison group, all of whom had some form of brain damage but still fell within the average normal curve of intelligence. Each participant was shown a series of faces portraying different emotions and asked to match each face with the proper

* A 1904 short story about a golf trip gone bad. A skeptical professor encounters a questionable artifact, engages with it in a way no believer of the supernatural ever would, and faces the consequences, including an encounter with a literal bedsheet ghost.

adjective. Neutral, happy, surprised, angry, disgusted, sad—every participant successfully identified all these expressions. But when it came to images showing the facial signature of fear—eyes open wide, eyebrows raised, nostrils flared—only SM scored poorly on identifying them as being scared.

Flashing pictures of faces was just the first step. Trying to answer the question "Does she even feel fear?," experimenters tried everything to shake SM. They showed her a compilation reel of video clips from classic horror and suspense films: a giant spider descending slowly onto an unaware victim alone in the shower (*Arachnophobia*), a close-up of a presumed dead man at a crime scene who suddenly sputters to life (*Se7en*), a knife-wielding man wearing a rubber mask chasing his victim (*Halloween*), and a ghost climbing out of a well (*The Ring*). As with the facial expression test, showing control video clips was necessary. The fear-provoking clips were intercalated with other clips intended to stir a different set of emotions, namely, disgust (*Pink Flamingos*: a person swoops a piece of fresh dog feces into their mouth, chews, smiles, and suppresses a gag before spitting it out), sadness (*Faces of Death*: actual footage of severely malnourished adults and children), and happiness (*America's Funniest Home Videos*: a compilation of babies laughing). She rated the disgusting, sad, and happy things as being an 8 out of 8, just like some of the control participants did. However, she never rated any of the scary scenes anything higher than a 1.

Sure, video stimuli can get our sympathetic nervous systems activated. At their root, however, scary movies are a way to enjoy the thrill of the rush without any of the risk since we know that none of those fear-provoking monsters can climb through the screen and attack us.* So the research team ramped up their efforts, trying to provoke SM with something a bit more concrete. They took SM and four strangers through the haunted house exhibit at

* *The Ring* being the only possible exception.

the notorious Waverly Hills Sanatorium* in Louisville, Kentucky. The others shrieked, jumped in surprise, and lagged behind out of apprehension. But SM simply smiled and laughed at the "monsters" hiding in the darkened hallways, ironically even startling one of the scare actors when she reached out to touch their mask.

Then the experimenters tried something more hands on. SM claimed to hate snakes and spiders. Naturally, when the research team brought her to an exotic pet store, they were puzzled by her behavior. She asked a total of 15 times if she could touch the largest and most dangerous snake in the store despite warnings from the employees that it wouldn't be safe. If it weren't for their protective watch, SM would've grabbed them out of pure curiosity. Even with respect to real-life stimuli, SM stood tall in the face of fear. Recalling a past incident when she was 30, she described being grabbed by a stranger on a park bench, being held at knifepoint, and then daring the assailant to try to cut her. Uninjured, she reports walking home calmly, never calling the police about the incident and passing by the same park the next day.

Experimenters, like explorers pushing into the dark vistas of the unknown, turned the intensity dial up in an attempt to find her limit. In one experiment, taken straight out of a medical torture film, SM and control participants were forced to breathe through a face mask supplied with air containing 35 percent carbon dioxide. For reference, normal ambient air contains such low levels of carbon dioxide that it is usually reported in parts per million, somewhere around 0.04 percent. This sensation is terrifying. It's like being buried alive. Within seconds, your brain, sensing high levels of carbon dioxide in the blood, feels suffocated, like you've been underwater for too long as you swim desperately to the surface. You gasp for air. You scream for help. You claw at the

* The former tuberculosis treatment center is believed to be a hot spot of paranormal activity, receiving its own episode of *Ghost Hunters* in 2006.

mask over your face. Surprisingly, this was also SM's reaction. The woman who reported no panic when threatened with a knife to her neck, for the first time, showed the experimenters that she could feel fear or, at the very least, the physiological signs of sympathetic nervous system activity.

As if mock suffocation wasn't enough, I find the next phase of the experiment to be particularly cruel. Participants were given a moment of normal air to reoxygenate their blood and mentally recover from the panic. Then they watched as the experimenters busied themselves with the equipment, just like they did previously before the first exposure to the high carbon dioxide air. Whereas Pavlov's dogs salivated at the sound of the bell that came before their meals,* these people were quickly conditioned to prepare for what came next: another round of suffocation torture. For most people, seeing these researchers dial knobs and adjust the gas tanks would invoke the classic fear response in anticipation of what is about to follow. But not in SM. Her heart rate and skin conductance response (a measure of arousal) remained unchanged. This experiment revealed a limitation of the amygdala's function. These circuits are important for the processing of external fears and launching the appropriate physiological response but don't have much to do with internal states, even if the downstream effect on the body is the same.

It wasn't that SM didn't understand the concept of fear. She demonstrated knowledge of the semantic content of words such as "terror," "frightened," and "scared," using them correctly in sentences. When listening to people speak using different tones of voice, she could identify which was supposed to be a scared voice.

* In a surprisingly heated debate about the origin of this experiment that appears in every introductory psychology textbook, there is a suggestion that "bell" is a mistranslation of "buzzer," which may have been the actual conditioned stimulus that Pavlov's research team used. Regardless of the sound, the appearance of the scientist who rang the bell or pushed the buzzer and fed the dogs would be an even stronger cue to initiate their salivation response.

She even experienced fear once as a young girl, presumably before the disease damaged enough of her amygdala, when a threatening Doberman pinscher cornered her. She comprehends and understands fear. She's just immune to its effects on her body.

Patient SM is a rare exception, living a life without caution. For the rest of us with our amygdalae, fear protects us against the terrors of the unknown in our often-scary world. It is this sense of fright at the right time that keeps us safe from predators, ghosts, and other spooky things that go bump in the night.

The first stroke of the razorlike crescent . . .

Edgar Allan Poe describes intense psychological horror in *The Pit and the Pendulum* when the protagonist first meets the razor-sharp blade descending slowly, closer and closer to the restrained narrator's chest. Franz Kafka invents a terrible execution device in his 1914 short story *In the Penal Colony*: the laws broken by the condemned were painfully inscribed into the victim's back over a 12-hour period. Yadin Dudai straps people down into a noisy coffin-sized tube for an hour while a meter-and-a-half-long snake* is brought closer to the person's head. But there is at least one major difference between Dudai and the likes of Poe and Kafka. Dudai isn't a master of horror imagery or one of Ian Fleming's scheming mega-villains prying secrets out of a British special agent. Dr. Dudai is a real person, subjecting other real people to his unique fear-inducing condition using the described apparatus, an fMRI machine in conjunction with a living snake on a conveyor belt. For many participants, these experiments are on a similar plane of psychological horror imagined by Poe or Kafka.

* Yes, again with the snakes.

The fMRI has been a great ally in neuroscience. It uses a combination of a powerful electromagnet* and a radio wave generator to noninvasively measure different particles in the brain, including oxygenated blood. A person in the scanner can perform a variety of behaviors, such as straightforward mentalization scenarios ("imagine you find a hundred-dollar bill on the ground") and more complex interactive tasks that require high-order processes that use both mental acuity and manual dexterity ("navigate a three-dimensional maze using this video game controller"). The theory behind fMRI is that whenever a person uses a specific circuit in their brain, the blood vessels that feed those areas dilate to deliver more oxygenated blood locally. That change in circulation can be detected by the fMRI.

Briefly put yourself in the shoes of those brave participants who volunteered their precious sanity in Dudai's 2010 experiment. The fMRI procedure can be a little bit challenging even without the snake. For one, the machine is really loud. Both the magnet and the radio wave emitter must spin rapidly around the participant, making a constant mechanical droning noise interrupted by periodic clicks and bangs. Then there is the claustrophobia. The scanning tube is around two feet in diameter, so the top of the machine is just inches away from your face. Technically, you aren't completely restrained like Poe's or Kafka's victims. But your head might be put into a restraining harness to help keep your head still; otherwise, any movement would produce blurry images. And if this confinement doesn't trigger a panic attack, Dudai added the most unusual stressor, a snake on a conveyor belt.

Even before this experiment, fMRI has reaffirmed the observations made in Patient SM regarding the link between the amygdala

* A typical clinical fMRI has a strength of 1.5 Tesla, which is stronger than a junkyard magnet used to lift cars. These magnets are so powerful that metallic objects in the room become literal missiles—at least one person has died because of an oxygen tank being accelerated into the scanning tunnel.

and fear. The amygdala fMRI signal ramps up whenever someone is fearful, whether because they are looking at pictures of spiders, controlling a digital avatar pursued by a virtual predator accompanied by a real electric shock if they get caught, or watching two of the top 10 scariest movies according to Rotten Tomatoes (even with a quantification of the number of jump scares). And amygdala activity decreases under brighter light conditions, answering the question that Nickelodeon posed many years ago: yes, we are Afraid of the Dark.

The participants in Dudai's experiment were given a simple instruction: Bring the snake as close to your head as possible. To do this, control over the snake's position on the conveyor belt was given over to the person in the scanner, moving closer to or farther away at the push of a button. The serpent in question was a corn snake, physically harmless to a grown adult. But that doesn't mean it can't cause psychological distress. Some participants were put into the experimental "fearful" group based on their responses to questions such as "I wouldn't take a course like biology if I had to dissect a snake" or "I would prefer not to finish a story if snakes were introduced to the plot." Another group of participants, the "fearless" group, had prior experience handling snakes, making them an excellent control for comparison.

Dudai was interested in a research question beyond fear. He wanted to understand the circuits involved in *overcoming* fear. This experimental setup gave these participants an opportunity to display bravery while observing brain changes. Whenever members of the "fearful" group brought the snake closer to them, the signal in the subgenual anterior cingulate cortex increased. The implication here is that neural signals running through this little patch of gray matter are involved anytime a person puts their fears aside and does something risky, whether it be to bury an ax in the head of a 40-foot-long anaconda, improvise a flamethrower out of an aerosol can and lighter to roast a nest of spiders, or chase a career in academia.

There was a clown in the stormdrain. —Stephen King, *It*

A simplified representation of a smiling face, such as an emoji, does not usually inspire an uneasy feeling. Neither does a photograph of a person smiling. But when those features of a normal happy face are exaggerated or distorted, perhaps by bright red strokes painted broadly over the lips or a face whitened by powdery makeup, that emotion can look less than human. Something vaguely normal without that undefinable something that makes it genuine can be quite horrific. Just think of the wide toothy grin of *Alice in Wonderland's* Cheshire Cat, Jack Nicholson's look of crazed madness as he smashes through the bathroom door in *The Shining*, and all those creepy smiles from, well, *Smile*. This experience applies to much more than just faces: the staggered limping of George Romero's zombies, the jerky mechanical movement of cinematic androids, or humanoids who nimbly crawl on all fours, such as the possessed Regan from *The Exorcist*, all evoke a sense of discomfort through a phenomenon called the uncanny valley effect.

Even though the phenomenon has been strongly replicated in human psychology, the uncanny valley was first described in 1970 by a robotics professor named Dr. Masahiro Mori. Mori plotted a theoretical graph that measures human-like characteristics on the horizontal axis and our affinity toward that robot on the vertical axis. As the robot gains a few traits that are more anthropomorphic, our affinity toward it increases. However, there is a certain point where there are too many human-like traits without fully achieving human-hood. And at this point, our affinity toward the robot dips dramatically. This low point of the graph is the uncanny valley.

It may be more useful to describe this graph as having four unique domains:

1. At the far-left end are robots with no human traits. These do not elicit any emotional response. We don't have strong reactions about electric can openers or automated vacuum robots. They look and behave nothing like humans, and we therefore don't feel any emotions toward them other than perhaps gratitude for making our lives just a little more convenient. They are perceived as a tool, not a friend or a foe.

2. As the robot starts to possess a few more humanoid characteristics, we begin to have more positive feelings toward them, so the curve rises toward the positive axis. There are two subdomains within this rising phase of the graph.

   Disney Pixar's *WALL-E* and *Big Hero 6*'s Baymax are robots on the left edge of this phase of the graph. Although WALL-E has wheels and Baymax is essentially a balloon, they possess a couple physical traits we see in ourselves. Both have two arms that they use to manipulate their surroundings. They can express a wide variety of human emotions using only language, ranging from wistful longing to happiness. With just a few anthropomorphic traits, we find WALL-E and Baymax cute and likable.

   An example of the right end of this domain is C-3PO, the golden robot from the *Star Wars* franchise. Much more human in appearance than the other two robots, "Threepio" has profoundly anthropomorphic traits: two metallic arms and legs and a head in which are set two circular lights for eyes, all attached to a central torso, giving it an average height and build. The proportions are human enough that actor Anthony Daniels was able to play the role of C-3PO in the earliest films by simply wearing pieces of the costume. Most of us find the robot endearing. Some of us have even empathized with it, feeling a passing twinge of sadness when a Tatooine bartender gruffly denies him entry: "No droids allowed!"*

3. This is the uncanny valley, the critical region of the graph in which the masters of horror thrive. Robots in this zone are more human-like than C-3PO, probably with skin and facial features. There are some tiny features about these robots that are just a little off. Maybe the way they blink is a fraction of a second slower than expected, like the actuators can't move as fast as a naturalistic blink. Or maybe they don't blink at all, producing an equally bizarre feeling. Maybe the transition from a neutral expression to a smile is just a bit too angular, as if the mechanical parts pulling at the lips are not synchronized properly with those around the eyes. Many times, it's very difficult to put your finger on exactly what feature is inhuman. Nonetheless, you experience intense discomfort when you see it. Ironically, being vaguely anthropomorphic without accurately portraying that critical element that makes us human is less appealing than having almost no human traits.
4. At the far-right end of the graph are other humans, who elicit the strongest empathic response. When we see another human in pain, we sympathize with them and feel their suffering.

The Actroid was one of the earliest humanoid robots, unveiled in 2003 at a Japanese robotics exhibition. With 40 or so different points of articulation, it had movement patterns that resembled

* My personal affinity for these robots had to be kept in check during writing. I had a predisposition toward anthropomorphizing these droids, having originally given both WALL-E and C3PO male pronouns when they are clearly nonliving.

natural biological functions, such as tiny shifts in head position during speaking and the rhythmic rising and falling of the chest. Its face was designed to look like the typical Japanese everywoman, created by averaging several real Japanese people. It seamlessly integrated floor sensors and omnidirectional vision sensors to maintain eye contact during conversation. It had a built-in AI chip that allowed it to function like a real human interviewer, listening to voices and responding appropriately, even carrying out simple small talk by asking questions such as "Where are you from?" Roboticists were amazed at the realism of the Actroid.

But the rest of the world saw something else. Instead of marveling at the skin-like characteristics of the face, we felt the Actroid's empty, soulless eyes. We saw the rubbery texture of silicone distorting in strange ways and the unnatural quickness with which the head rotated and angled. We noted that brief, half-second-long delay between each line of conversation as the robot ran through calculations about how to respond to even the simplest questions before responding coldly. We experienced the depths of the uncanny valley firsthand. Even tech website CNET, often optimistic about all things future, describes later versions of the Actroid as creepy.

We even see the effects of the uncanny valley in two-dimensional characters on-screen. The characters from holiday film *The Polar Express* were called "blank-eyed and rubbery-looking as moving mannequins—the stuff of nightmares, not dreams." In 2001, the first fully computer-animated film using motion capture and photorealistic characters, *Final Fantasy: The Spirits Within*, was a box office flop, partially due to the visuals being "more discomfiting than appealing." That same year, DreamWorks designed characters from their upcoming animated film *Shrek*, including a human love interest, Princess Fiona. On showing her original design to a screener group of children, some of them started to cry, prompting the animators to push her design more toward the cartoony rather than the realistic. Even the best animators in the world can't escape the uncanny valley.

Of course, directors of horror films have used this phenomenon to great effect, playing with our intrinsic fear of the distorted human. Michael Myers, the iconic villain from *Halloween*, wears an unmoving and featureless mask, adding to the mystery of the relentless killer. M3GAN's slightly too large eyes and calculating manner of speech gives more Actroid vibes than childhood best friend/toy. Sadako, the ghost who climbs out of the well in *The Ring* (1998), was filmed walking backward, then that footage was reversed to put a little more weird in her walk. Each member of the doppelgänger family from *Us* had something uncanny going on, between Abraham's grunts, Pluto's crawl walk, and Red's gravelly voice modeled after a genuine medical condition.

One big question remains: Why are we wired to feel discomfort toward the uncanny? Danger avoidance theory offers an evolution-based reason, suggesting that uncanny stimuli provoke an extension of disgust. Physical disgust makes us queasy in response to rotten food or the smell of decay since we benefit from keeping away from things that could make us ill. Danger avoidance theory suggests that visual stimuli can also initiate a disgust response, causing us to fear and therefore avoid the uncanny. For example, compare the intensity of the emotional reaction you would feel on encountering the three following conditions:

1. A dead ant. Upturned with their tiny legs curled inward, the unmoving body of an ant poses almost no imminent threat to one's safety. Most of us have zero emotional reaction to the bug. Realistically, it was our very own finger or boot that was the cause of that ant's death.
2. A dead squirrel. The emotional response becomes more pronounced when we see a body of a small mammal. Compared

to insects, there's a bit more substance to mammals, more flesh on the animal. We think more about the circumstances that led to their demise. Could there be a coyote or other predator nearby that could also threaten my own safety? Seeing the squirrel's insides on the outside is cause for concern. Could a zoonotic disease agent or the cloud of flies buzzing around the animal pass an illness on to me? Either way, the sight elicits an appropriate healthy nausea causing us to want to get away.

3. A human corpse. Obviously, the most distressing sight. I bet most people remember the first time they came across an unexpected corpse. For me, his leg was twisted unnaturally at the knee, face down in an alleyway late at night one evening. I barely slept.

A dead body sits at the depths of the uncanny valley. Cadavers lack even the subtlest signs of life. Blinking stops, the rhythmic ups and downs of breathing are gone, and the skin appears blanched. We pick up on the absence of those features very quickly. Just to the right of the deepest point are living people but perhaps those who are actively fighting some kind of illness. Those basic signs of life are still there at a rudimentary level, but they may appear weakened. Maybe it looks like slowed responses or motor activity, shallow breathing, and pallid or clammy skin. Both corpses and ill-appearing people can produce an avoidance reaction according to danger avoidance theory. We sense that something is wrong out of some evolutionary protective sense that a nearby threat could turn us into a second corpse or that the person could pass some infectious agent on to us.

Another possible explanation for our "ick" response has to do with mortality salience, a shift in mindset that happens whenever we are reminded of our own inevitable deaths. Uncanny stimuli are that reminder. Think of the Pale Man, the second-most-horrific entity in *Pan's Labyrinth*.* Seated silently at the head of an overabundant banquet table, his pallid gray skin hangs loosely over a skeletal frame. He reminds us of our own frailty and physical weakness when we reach the end of life. An android with limbs that are just a little too stiff or a face that shows little emotion reminds us that we may lose control over our bodies as we age, matching the symptoms of terminal illnesses, such as muscle rigidity in Parkinson's disease and flattened affect in Alzheimer's disease dementia.

All these behavioral responses have some underlying neurobiological substrates. The ventromedial prefrontal cortex (vmPFC), found at the base of the frontal lobe, is key to our uncanny valley experience. Imaging studies suggest that signals in this area decrease when we are asked to rate the likability of uncanny residents, such as artificial humans modeled after people who had received extreme plastic surgery photographed under bizarre lighting conditions with reduced contrast to make their skin appear gray. The vmPFC communicates strongly with the amygdala, potentially functioning as a clamp that inhibits amygdala activity, thereby acting as a repressor of our fear circuits. As a subregion of the PFC, it contributes to decision making. When faced with whatever creepy androids are standing just around the corner, these circuits help us decide that they should be avoided.

The insula is likely implicated as well, that often-forgotten lobe of the brain that is involved with initiating the emotion of disgust. When shown a continuum of faces ranging from computer-generated human face avatars to actual photographs of human faces, those intermediate, sort-of-human-but-not-quite stimuli provoke an insula response. You could imagine that eating something rotten initiates physical disgust. From a self-preservation

* The worst being the senselessly violent fascist dictatorship that ruled over the country in the wake of the Spanish Civil War, the backdrop against which Guillermo del Toro's dark escapist fairy tale is set.

point of view, that same feeling could also be appropriate when encountering some uncanny stimulus, which may have many traits in common with a cadaver.

> Over the trees, occasionally, between them and the hills, she caught glimpses of what must be the roofs, perhaps a tower, of Hill House. They made houses so oddly back when Hill House was built, she thought; they put towers and turrets and buttresses and wooden lace on them, even sometimes Gothic spires and gargoyles; nothing was ever left undecorated. —Shirley Jackson, *The Haunting of Hill House**

A lone wheelchair placed in the middle of an abandoned hospital hallway. A back room of a mall department store, populated only by headless and armless mannequins. An amusement park, equipment powered down for the night. All these scenes inspire a unique kind of tension if you let your imagination run a little. Why are these places empty? Where did all the people go? More ominously, what happened to the people? Even images of a house for sale, completely devoid of furniture with only subtle divots in the carpet to suggest that anyone had ever lived there, give off an eerie vibe.

For a writer or filmmaker, depicting a living thing as scary can be straightforward. Give the alien some teeth, claws, and a third, fourth, or seventieth eye, and you'll freak people out. Give the extra-dimensional beings and elder gods some motivations beyond our understanding, and the audience will read them as chaotic and dangerous. But places and settings can be equally spooky, a notion that classic gothic horror authors have played with for centuries. Shirley Jackson's Hill House, that abandoned mansion full of weird

* In reference to an earlier footnote, *The Haunting of Hill House* (1959) is Stephen King's other pick for finest supernatural horror story. Here, a professor invites some people to stay in a haunted house where they find what they are looking for, but no one is happy about it.

happenings, is arguably just as much a character of her book as those living, breathing investigators who enter through those doors.

The odd feeling of being in some transitory time or space, such as a back room or a house for sale, has come to be called liminal space horror. We have all felt liminal horror, especially around March 2020 during the peak of the pandemic. Think back to the first time you saw an image of Times Square without any cars, the pope delivering a blessing before an empty St. Peter's Square, or an open-air market in Asia without the usual throngs of people.

A classic example of a liminal space horror in film is the depiction of the Overlook Hotel in Stanley Kubrick's film adaptation of *The Shining* (1980). Of course, the iconic twins and the "Here's Johnny!" scene are haunting. But between the physically impossible layout of the hotel, the disorienting chase through the hedge maze, and exploration of the hallways normally reserved for the waitstaff, the setting itself has just as much character and nuance as the people on-screen. In more recent media, *Five Nights at Freddy's* (2023) evokes that same feeling. The survival horror video game franchise that inspired the movie puts you in the shoes of an overnight security guard watching a children's entertainment pizza venue, normally full of life, rowdy with excited children and drunk parents. But at night, the animatronics break free of their programming loop. If you don't manage your resources properly, you are punished with a brutal jump scare as they pop out of the air ducts. Both the games and the films play into the fear of why abandoned places are so eerie.

What's so powerful about liminal horror is that it doesn't present the classic environmental cues that signal scary places. Obviously, dark, claustrophobic tunnels of unexplored cave systems and forgotten spacecraft hide terrible secrets. But even brightly lit venues and open spaces can be used effectively in liminal horror. Consider the overly vibrant colors of the Overlook Hotel's carpet, the tranquil countryside of *A Quiet Place* (2018), or the rows of empty suburban homes in postapocalyptic settings. In a way, liminal spaces are analogous to the uncanny valley but for physical surroundings. There's just something unsettling about places that should be full of people when they aren't, and that sets off our fear sensation.

Despite the many examples of liminal horror throughout fiction storytelling, the underlying science of why these places are scary is not well understood. Part of it begins with the scene-perception network, a series of cortical areas that are activated when we are trying to identify our surroundings. These circuits were probably critical for successful wayfinding to and from sources of food or homes, as our ancestors relied on visual cues to navigate across the plains. This network is made up of at least three structures.

1. The parahippocampal place area (PPA), located roughly at the base of the brain, responds fairly selectively for images of houses and buildings compared to faces or objects. That signal is even stronger when looking at rooms and outdoor scenery, both constructed, such as "The Bean" at Millennium Park in downtown Chicago, and natural, such as a cascading mountain range. The signal here fires up even when the room is empty, suggesting that these circuits encode our surroundings with respect to physical boundaries.
2. The retrosplenial complex (RSC) increases its signal to similar stimuli. Interestingly, fMRI signals of both the PPA and the RSC rise even when a blind person is using their hands to explore a physical space, such as a miniature room built out of Lego bricks. These brain structures really care about spaces and your place in them, not just about the sense of sight.
3. The occipital place area (OPA) is the posteriormost of this triad and the least well studied. It likely also encodes bound-

> ary information and has a strong role in identifying individual objects within the surroundings rather than processing a more holistic view.

It's possible that some frontal areas are functionally connected with this scene-perception network. In the same way that the vmPFC signal dips when seeing uncanny people, it might likewise release the clamp on amygdala-related fear behaviors in response to liminal spaces. It is similarly likely that downstream of the scene-perception network includes insula structures. A change of activity along this pathway might produce that slight "something is off, I'm not supposed to be here" sense of danger feeling rather than outright fear.

There are these circuits in place that keep us a little wary of liminal spaces such as *The Twilight Zone*, which, as Rod Serling says, exists in the "middle ground between light and shadow." Liminal spaces and the feelings they provoke might be one of the thrills that drives urban explorers to crawl through decrepit service tunnels, the rationale for why Pompeii evolved into a tourist destination, and the raison d'être for anyone in the academic discipline of archaeology. But people override that preservation instinct whenever a hint of curiosity creeps up on those scientist-explorers who push past their hesitation and continue to probe into the unknown. As we will see over the next chapter, just because you can . . .

. . . well, maybe you still should anyway.

CHAPTER TWO

"THE SPARK OF BEING"

Life, Death, and Reanimation

Experimental biology is far more gruesome than glamorous. The best way to study something, living or dead, is to take it apart. Usually, that involves nasty verbs, such as "cut," "crack," and "pull." While more sterile, knife technology hasn't changed much since the days of medieval barber surgeons. At the heart of it all, biologists often find themselves in laboratories surrounded by body parts that aren't their own.

For this fact, Mary Shelley's depiction of her titular scientist Victor Frankenstein, furtively stitching together limbs of the deceased in his secret workshop, is a reasonable approximation of experimental biologists in the 1800s. These were the scientists who laid the foundation for modern neurobiology. One name, Luigi Galvani, had been so influential among the educated high society that Shelley even explicitly acknowledged him in her seminal work. There is no doubt that his experiments inspired Shelley in 1816, that "year without a summer." He had shown that electricity and animal movement were closely related, a theme that Shelley explored more fully in her work.

> It was on a dreary night of November that I beheld the accomplishment of my toils. With an anxiety that almost amounted to agony, I collected the instruments of life around me, that I might infuse a spark of being into the lifeless thing that lay at my feet.
> —Mary Shelley, *Frankenstein*

But before we get to the story of Galvani's gruesome but groundbreaking discoveries, it might be helpful to put his observation in the context of the previous beliefs about the science of movement. From the days of the accomplished Greek physician Galen in 100 CE up until the 1500s, balloonist theory dominated the discourse. Dissections had revealed that thin strands of fibers, nerves, were physically connected to muscles. Balloonists theorized that nerves were hollow vessels that served as channels through which air or fluid can travel, allowing muscles to get filled up with that substance, leading to contraction. This theory went unquestioned for more than a millennium.

Then the Enlightenment happened. With it, the scientific method rose to prominence. Evidence in the form of astute observations and carefully planned experiments was the key to creating new knowledge. The Dutch microscopist Jan Swammerdam and his rigorous science discounted balloonist theories. Swammerdam dedicated much of his energy to dissecting and documenting the anatomy of frogs. In the 1660s, he developed a preparation consisting of an airtight glass vacuum tube with a long, narrow tube into which a droplet of water was placed. A frog leg, surgically deconstructed and stripped away to just bones, muscles, and nerves, was placed inside the vacuum tube. If there was a change in muscle volume on contraction, as a balloonist would predict, the small droplet of water would be displaced out of the tube. But the droplet remained still, indicating that muscles did not inflate during contraction. In this single well-designed experiment, he buried the balloonist theory of movement, putting hundreds of years of dogma underground.

A century later, Galvani would develop a similar frog leg preparation that allowed him to flesh out a more complete theory of animal movement, one that addresses the why. The romantic ideal of the serendipitous scientist stumbling into earthshaking discovery is often played up for dramatic effect, and Galvani isn't immune to folklore. According to the legend, he unintentionally discovered the mechanism with a little bit of static electricity.

Galvani's experimental preparation was like Swammerdam's minus the vacuum tube. The legs of the frog were dissected out and separated from the body. Critically, the internal "stuff" is kept intact at the level of the torso, including muscles, bones, tendons, and, most important, the nerves. Initially, Galvani delivered a small jolt of electricity to the slimy frog skin and was surprised when the muscle remained still. No contraction of the leg muscle, not even a twitch. The popular myth suggests that an accidental circuit through some nearby electrical equipment and the metal scalpel he was holding sent a tiny zap of electricity through the intact nerve, provoking a strong muscle contraction as if the headless frog were still alive. This muscle action led Galvani to theorize that some kind of electrical activity was responsible for movement in living animals. Naturally, these shocking results sparked an interest in the intersection of biology and electricity. In 1786, his theory of movement, called Galvanism, was born.

Galvani's nephew, Giovanni Aldini, continued in his uncle's footsteps. He took these scientific advancements in bioelectricity out of the laboratory and into the public, attracting audiences the same way a modern field museum or aquarium might. He had experimented with electrical nerve stimulation on a whole range of dead animals: decapitated oxen, horses, sheep, and dogs were all given "life" by electrically stimulating exposed nerves. Despite the gruesome nature of these public displays of bioelectricity in action, they became very popular, bolstered by Aldini's natural talent for showmanship. The once-lifeless heads would contort wildly, gnashing the teeth and opening the eyes terrifyingly. The headless bodies thrashed recklessly on the table, as if in pain. It was horrific. It defied nature. It stood against the Christian God. And audiences loved it.

> In one triumphant demonstration West was about to relegate the mystery of life to the category of myth. The body now twitched more vigorously, and beneath our avid eyes commenced to heave in a frightful way. The arms stirred disquietingly, the legs drew up, and various muscles contracted in a repulsive kind of writhing. —H. P. Lovecraft, *Herbert West-Reanimator*

In 1751 England, Parliament passed the Murder Act, a law introducing disincentives against those found guilty of homicide (apparently "execution by hanging until death" wasn't disincentive enough). Under the new provisions, executed criminals were denied a proper burial. Instead, their corpses were to be either displayed publicly or dissected to "better prevent . . . the horrid crime of murder," much to the delight of physicians and medical students around England who were buying corpses dug out of the ground to fulfill their demand of cadavers to practice on. These provisions permitted Aldini to push his experimentation beyond shocking common farm animals. Suddenly, there was a steady supply of bodies from which Aldini could "recruit" his new test subjects. Were H. P. Lovecraft's maniacal surgeon-gone-mad Herbert West a real citizen of 1750s England, you can bet that he and Aldini would be standing beneath the gallows, elbowing each other to get first dibs on the freshest state-sanctioned cadavers.

George Forster of London was Aldini's most famous subject. Found guilty of murdering his wife and child by drowning them in the Paddington Canal, Forster was condemned to hang until death. Shortly after Forster's execution, with the approval of the government, Aldini went straight to work. He already knew that when given the right type of electrical stimulus in the right place, dead animals can move as if they were alive. Since man is just another animal, why not demonstrate that electricity can reanimate dead humans as well? Aldini placed two large metal rods connected to a powerful battery in contact with different parts of Forster's lifeless corpse. Once activated, the massive current passing through the dead body triggered bizarre combinations of muscle activity, causing Forster's corpse to jerk violently.

"On the first application of the process to the face, the jaws of the deceased criminal began to quiver, and the adjoining muscles were horribly contorted, and one eye was actually opened. In the subsequent part of the process the right hand was raised and clenched, and the legs and thighs were set in motion."

Aldini's displays of Galvanism aren't a far cry from the work of Dr. Frankenstein. Compare with Shelley's words: "By the glimmer of the half-extinguished light, I saw the dull yellow eye of the creature open; it breathed hard, and a convulsive motion agitated its limbs."

A later description of neuronal electrosensitivity confirmed what Galvani and Aldini discovered in the intact organism. These experiments were done, strangely enough, in squid. Far from the monstrous tentacled horrors of Lovecraft's imagination, you are more likely to eat a *Loligo forbesii* than the other way around. In the context of neuroscience, squid have one especially unique feature that makes them ideal for dissection in the laboratory. Squid anatomists discovered a large hollow structure that was believed to carry blood to innervate the squid's main escape mechanism, a water propulsion reflex used in juking predators. This tube was surprisingly the axon of a single neuron. Axons are the output structure of the neuron, and our nerves are made up of a bundle of

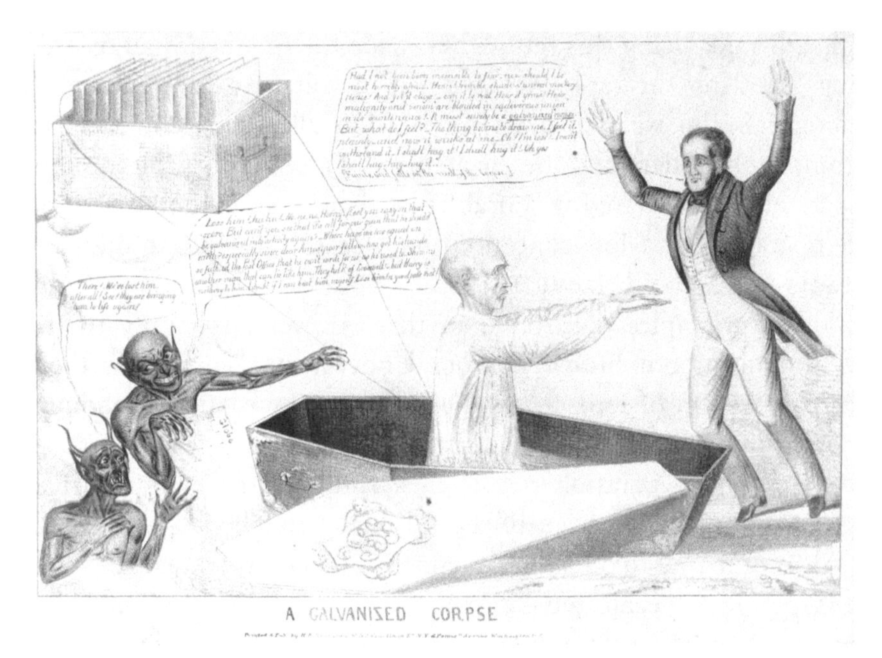

Figure 2.1. Two devils, one of them expressing concern that their Dark Lord no longer has domain over Forster's soul due to Aldini's meddling.
HENRY ROBINSON, PUBLIC DOMAIN, VIA WIKIMEDIA COMMONS

axons. Although human axons may be unusually long, they are not very wide, with a diameter typically around 25 microns (roughly 0.001 inches). But in the squid, this giant axon is half a millimeter—nearly 20 times wider.

A huge cell provides plenty of advantages, especially from an experimentalist's perspective. Two researchers, Alan Hodgkin and Andrew Huxley,* had the idea to dissect out a small section of the squid's giant axon for study in artificial, well-controlled conditions. Into one end of the cell, they threaded a hollow glass pipette filled with briny seawater and a thin silver wire. The other end of the cell was connected to an electric source that they could use to stimulate the cell, as if it were receiving signals from an

* Yes, he's related: he's the half brother of Aldous Huxley, author of *Brave New World*.

intact squid. Using this setup, they made the first neuronal recordings showing the rapid rise and fall in voltage during an action potential. The isolated axon preparation allowed them to more completely understand the mechanism of communication through systematically changing variables such as stimulation strength, temperature, and ion concentration. For this work, the duo was awarded a Nobel Prize in 1963.

The principles discovered in these nerves also apply to the communication in human neurons. Under normal conditions, there is a separation of ions across the cell membrane, such as sodium, potassium, chloride, and calcium. Since these ions carry electrical charges, this separation results in a tiny difference in electrical potential, with the inside of the cell being roughly –70 millivolts (mV) compared to outside. When the cell's potential reaches some a threshold value, around –55 mV, the neuron does something surprising. In a self-sustaining process, the charge of the cell rockets to positive values, sometimes even reaching as high as +30 mV. Then, within a millisecond or two, the voltage falls back down, overshooting to around –80 mV before returning to the resting value of –70 mV. This sudden change in electrical potential is the action potential, that all-or-nothing signal that is the basis for almost* all neural communication.

> here, here!—it is the beating of his hideous heart! —Edgar Allan Poe, *The Tell-Tale Heart*

While Herbert West was injecting his mysterious reanimation cocktail in the pulp magazines of Lovecraft's 1921 story, research scientists were hard at work discovering the exact nature of the

* Some parts of our sensory systems, like the photoreceptors of our eyes and the hair cells of our inner ears, care more about graded potentials, which are small, subthreshold changes in membrane potentials.

chemicals used to signal to our internal organs. Among those inspired by Galvani's demonstration of bioelectricity was a German pharmacologist named Otto Loewi. Loewi posed this question: What if Galvani's bioelectricity theory was only the first part of the story and the change in electrical properties was just the first of many steps along the communication process?

A testament to the inherently bizarre nature of experimental biology, Loewi's experiment to answer this question could have been copied straight from the pages of Shelley, Poe, or Lovecraft. Loewi, the skilled neuroanatomist that he was, cut open and exposed a frog's thoracic cavity. Then he extracted the tiny, still-beating heart in pristine condition with the major nerves intact. He had known from previous research conducted in the 1890s that electrically stimulating a particular nerve called the vagus nerve would slow down the heart rate. The vagus nerve is the key driver of the parasympathetic nervous system, a branch of the nervous system responsible for the rest-and-relaxation response. This set of nerves is the counterpart of the sympathetic nervous system described previously, the one that kicks up when we feel fear. Instead of ramping up heart rate and respiration in preparation for combat, vagus nerve activation leads to slowed heart rate and breathing and decreased blood pressure. This slowing of physiological measures contributes to that feeling of intense sleepiness after a huge meal.*

Loewi needed an experiment to prove that chemicals released by nerves are the ultimate influencer of heart rate rather than electricity. The perfect experiment came to him in a dream or, rather, maybe a vivisection nightmare. The next morning, Loewi rushed to his laboratory as he tried to decode his late-night scrawling. He isolated two hearts from different frogs, each heart now beating

* This relaxation response is not entirely driven by the vagus nerve. Another set of nerves called the enteric nervous system tells the gastrointestinal system how to act, and these nerves function even if the vagus nerve is severed.

independently in their own saltwater bath. After stimulating the vagus nerve of the first heart and observing the expected slowing of heart rate, Loewi collected some of the solution in which the first heart was bathed. When this solution was washed on top of the second heart, it also slowed despite not being electrically connected to the first. Armed with these results, Loewi could back up his claim that chemicals were released by the electrical impulses sent down the vagus nerve, and it is ultimately this chemical, not the electrical stimulation, that causes the heart to slow. He named this chemical signal Vagusstoff. A decade or so later, Loewi shared the 1936 Nobel Prize with his lifelong friend and chemist Sir Henry Dale for the discovery and characterization of the molecule that came to be called acetylcholine. Today, we know that acetylcholine is conserved across many living things, ranging from bamboo plants and fungi, all the way through sea sponges, insects, frogs, and humans. Acetylcholine is also the chemical that Galvani and Aldini manipulated with their "reanimation" experiments. Delivering an electrical stimulation to the nerves causes neuronal action potentials to fire, resulting in the release of acetylcholine at the end of the axon. Acetylcholine then travels across the neuromuscular junction, causing the frog leg to kick or Forster's body to convulse horribly.

Dale and Loewi's work was the first scientific description of a neurotransmitter: molecules produced by neurons that communicate to some other target cell. So far, more than 100 different neurotransmitters have been described, with dopamine, serotonin, and histamine being among the most well known. Much of the electrical chatter in the brain is neuron-to-neuron talk, where neurotransmitters change the electrical properties of other neurons. Some neurotransmitters, such as glutamate, are mostly excitatory, while others, such as GABA, are mostly inhibitory. Neurotransmitters can influence more than just neurons. As Loewi showed, the heart beats faster when norepinephrine acts and slower when acetylcholine takes over.

Confusingly, the same neurotransmitter can push cells in two completely opposite directions, adding another layer of complexity to the story. For example, acetylcholine slows down heart cells but excites muscles. This paradox is accounted for by different types of receptors that are found on different tissues. Receptors are submicroscopic molecular machines found on the surface of cells. Neurotransmitter molecules temporarily bind to these receptors, and those receptors act as the intermediary between the exterior of the cell and what is going on inside. When acetylcholine binds to the muscarinic acetylcholine receptors on the heart, it tells the cell contraction rate to decrease. But when that same neurotransmitter binds to nicotinic acetylcholine receptors on the muscle, those receptors physically open up a pore, allowing ions such as sodium to cross the membrane of the cell, eventually leading to contraction.

That separation of ions allows for both muscle activity and neuronal communication. Ions having different concentrations outside compared to inside cells is a state of low entropy that requires constant energy expenditure to enforce that kind of order. Within hours after death, the body stops producing ATP, the molecule that cells use for energy. As the supply of ATP dwindles, the processes by which ions are forcibly pushed across the neuronal membrane slow and eventually stop. The electrical difference between the inside and the outside of the cells shrinks from –70 mV to 0, thus preventing action potentials from firing. While Dr. Frankenstein doesn't concern himself much with the age of body parts he scavenges from those charnel houses and unhallowed graves, Dr. West is onto something in his comical obsession with getting the freshest bodies possible. No ATP means no action potentials, no receptor activation, no contraction, and no "spark of life" regardless of how much current you pump through the inanimate body.

Echoing Dr. West's ambitions, the need for fresh samples in experimental biology is absolutely real, and I've seen it firsthand. When I was studying neural communication in the mouse brain, easily more than half of the time in the laboratory was dedicated to getting the healthiest possible samples. The brain is kept submerged in an ice solution while a razor blade vibrates horizontally, slicing through the tissue to make quarter-of-a-millimeter-thin slices, roughly the thickness of an average index card. A perfect balance of salts and a heaping dose of sugars and ATP must be precisely measured until you fall within a window of acceptable osmolarity, which prevents the cells from shrinking from dehydration or swelling until they pop. The samples float in a warm bath of artificial cerebrospinal fluid, oxygenated via an aquarium bubbler. Finally, after hours of prep work and minutes of racing against inevitable cellular death during the dissection and extraction of the brain, the work can begin. Neurons are ready to be probed, stimulated, inhibited, and manipulated.

The longest I've ever kept a section of a brain alive in the laboratory was around six hours. At no point did the brain try to murder me.

> progress report 6-Mar 15. The operashun dint hurt. He did it while I was sleeping. They took off the bandijis from my head today so I can make a PROGRESS REPORT —Daniel Keyes, *Flowers for Algernon*

In 1967, Dr. Christiaan Barnard performed a horrifying operation. A team of bloody-gloved people, including his younger brother, would move around him quietly, disrupted only by the various machines that whirred and beeped rhythmically. In the early hours of December 3, Barnard stared into the vacant chest cavity of his first victim, whose sternum had been sawed vertically down the middle. His second victim went under the knife for five hours, completely unconscious and unaware of the surgeons delicately sewing together blood vessels. The first victim, a 25-year-old woman, died on the table. The second victim died 18 days later.

This isn't the opening scene of another horror story of medicine gone wrong or doctor gone mad. Quite the contrary. This is the story of the first human-to-human heart transplant.

The procedure was initially met with disgust. Instead of seeing the potential value of such a lifesaving endeavor, the public recoiled against the display of inhumanity. People from around the world wrote scathing letters, calling Barnard sadistic, abnormal, and a butcher. They called for him to be arrested. "Remove what?" "Cut where?" Operations carry risks, and when a new approach to a problem comes forward, even the most well-read and scientifically literate people may be quick to criticize. The heart transplant, like the allogeneic skin graft or the kidney transplant before it, were all considered wildly experimental and unnecessarily risky when first proposed.

But "radical" procedures of the past become routine. Surgical protocols get tweaked and optimized in light of new technology, each step improving alongside our rapidly growing body of knowledge regarding human physiology and biology. Given enough improvements in medicine and accompanying success stories, once-experimental surgeries such as the heart transplant become widely accepted: an improved version of Barnard's heart transplant procedure will be performed more than 4,000 times each year in the United States alone.

With the experience of highly skilled surgeons backed by the research of dedicated health scientists, nearly anything can be transplanted from one person to another: liver, pancreas, kidneys, lungs, sections of intestine, and patches of skin. Corneal grafts can help repair injured eyes, and the injection of someone else's bone marrow can help the body recover during cancer therapy. Even entire functioning units, such as a hand, can be taken from a donor and reattached to another person. The process of connecting bones, tendons, blood vessels, muscles, and nerves is challenging, and the hand transplant procedure has been performed only about 100 times since 1998.* To ensure that all the right connections are trained properly, hand recipients undergo intense physical rehabilitation, after which they usually recover enough function to manipulate their external environment in a rudimentary but functional way, restoring their ability to pick up a glass of water or tie shoelaces.

Despite these successes, some lifesaving transplant procedures are still in their infancy. And considering that more than 100,000 Americans are currently on an organ waiting list that grows every eight minutes as 17 people die each day waiting for an organ, finding a better system for organ transplants would improve several lives. A major breakthrough would be successful xenotransplantation, the transplant between species. Instead of waiting for a comatose, brain-dead organ donor such as Barnard's patient, a surgeon could harvest organs from a medical animal farm. There has been some precedent of xenotransplantation in the past, mostly with poor success. The first pig cornea implanted in a human was as far back as 1838, and it failed in two weeks. Of the 13 human recipients of chimpanzee kidneys in the 1960s, the longest survival duration was nine months. And for the hundreds of aging men who received sections of baboon testicles sliced into their own testes to restore their "zest for life," some of them experienced painful inflammation or infection.

One of the biggest problems with transplantation is interference by the immune system, that complex set of cellular and biochemical protective mechanisms that searches for foreign pathogens such as infectious bacteria or viruses and destroys them.

* Like Luke Skywalker, the first of these patients also lost his right hand at the level of the forearm. Unlike Luke, the wound wasn't instantly cauterized by the searing heat of a lightsaber: amputation by circular saw is a bloody mess.

Even in human-to-human transplants, once a transplant surgery is completed, the immune system recognizes the marker molecules on the transplanted tissue as not being a normal part of the body, thus targeting it for destruction.* To prevent this, patients take a strong regimen of immunosuppressant drugs, thus weakening the immune system just enough to prevent the body from rejecting the newly transplanted tissue. Managing the dosage is an important step in postoperative care. Too low, and the body mounts up inflammatory defenses and destroys the organ. Too high, and the patient becomes prone to developing infections. Case in point: Dr. Barnard's patient died from severe pneumonia, a consequence of a weakened immune system, not heart failure.

It is no surprise that nonhuman organs have molecular markers that are even more different than those on human organs. Overwhelmingly, this organ rejection was the main reason why past xenotransplants led to such poor outcomes. In a glimpse of what the future of organ transplant medicine might look like, a University of Maryland surgical team in January 2022 performed a human xenotransplant from a genetically modified pig. Using the many gene-editing tools available, these pigs were designed to have fewer "pig" markers and more "human" markers on the surface of their cells, thus allowing the pig heart to hide in plain sight from the immune system. Although the recipient, 57-year-old David Bennett, survived only two months with his new heart,† the surgical team demonstrated that this operation was at least feasible from the plumbing perspective since the blood vessels and nerves were wired the way they were supposed to be.

The next step in transplantation? How about extending a person's life by transplanting their brain into a new body? It turns out

* This problem was easily solved by the scientists in *Parts: The Clonus Horror* (1979) and *The Island* (2005).

† Some wonder if Bennett deserved this renewed gift of life, particularly the family of Edward Shumaker, the man whom Bennett stabbed seven times and left paralyzed in 1988.

that the wiring that connects the brain to the body is smaller, more intricate, and more numerous than that of any organ thus far transplanted. It would be much easier to just transplant the entire head.

To be more accurate, it's more like transplanting a new body onto a head, a radical surgery proposed by Italian neurosurgeon Dr. Sergio Canavero in a 2013 publication. This process has been dubbed "HEAVEN," or the HEad Anastomosis VENture procedure. On paper, Canavero's proposed surgery is a natural extension of hand transplant surgery. You are just moving a body part from one person to another. HEAVEN simply proposes putting an entire body on a head.

Canavero's willingness to perform such a risky medical procedure makes him a pioneer in surgery, navigating an as-of-yet-unexplored frontier filled with questions. Where would the cuts be made? How much pain would the disconnected head feel, and how would you keep it anesthetized? Would the surgeons be able to stitch together the tendons, muscles, blood vessels, and nerves fast enough before either the body or the head lose vital strength? How well will the new body, internal organs and muscles alike, take orders coming from a foreign brain? Would the body, which contains the overwhelming bulk of the immune system, perceive the head as a foreign invader and attempt to destroy it as if it were a virus?

Attaching a donor body to a recipient head introduces several challenges over simpler transplant surgeries. To address these challenges, Canavero and his international surgical team have developed a thoughtful strategy for the HEAVEN process. Most critically, they must maintain the flow of oxygen and nutrients to the living brain. In the intact body, the lungs, heart, and circulatory system are tasked with this essential function. Oxygenated blood courses through our blood vessels, which delivers oxygen to the metabolically hungry brain. But once separated from the body, the detached head essentially suffocates and quickly begins to die. By artificially pumping blood through the major vessels of the head to mimic the heart's main purpose, the surgical team hopes that the brain cells get the oxygen they need.

One of the biggest challenges of Canavero's endeavor will be the reattachment of the nerves, a notoriously difficult process. A Chinese orthopedic surgeon who was on-site during the first hand transplant, Dr. Xiaoping Ren, has been working diligently on this question and has made some major advances. Ren's spinal cord repair procedure uses an extremely sharp diamond blade to sever the nerves followed by treatment with a chemical agent to help those damaged cells re-form, something the team calls a fusogen. Treatment with polyethylene glycol (PEG) stabilized a transected piece of the spinal cord, essentially acting as cellular glue that mends the severed cell membranes. This restored motor control, allowing the subjects to move around as if they were uninjured. Furthermore, imaging strategies show that PEG restores the white matter connections between the brain and the extremities, a key indicator of success,* that is, if you can put aside the notion that Ren's list of experimental surgeries read like excerpts of Herbert West's diary, with entries such as "Today, I swapped two cadaver's heads" and "I sewed a rat head on top of another rat's head."

Other studies had questionable results. Some rats died in the process; many can breathe independently, but even these survivors don't live to see 24 hours. In one case, as if by the very hand of the Old Testament God, a group of experimentally injured PEG-treated rats died in a flood that destroyed a research lab in South Korea.

Compared to the medical questions, the ethical questions are even more numerous and far more difficult to answer. If the patient dies within the month, as Barnard's heart transplant patient did,

* PEG might be a chemical you've heard of in different contexts, depending on your career and hobbies. Medical professionals might recognize PEG as the main ingredient in laxatives. Archaeologists and history buffs might remember when PEG was used to preserve the original colors of the *Terra-Cotta Warriors*, ancient statues from Xian province in China. Others might recognize this slippery, viscous liquid as an ingredient in personal lubricants.

would Canavero be at fault and therefore subject to civil or criminal charges? What if the patient lives but has their mobility worsened and is now subject to frequent medical examinations, resulting in a significantly worse quality of life than before? What about the donor body? Where will that come from? Will this procedure be available for anyone with the cash?* Or will it be reserved only in the case of medical necessity? And what even defines "medical necessity"?

Critics were quick to mobilize in light of these philosophical quandaries. Canavero has been a target for bioethicists, who have called the head transplant an "unproven cruel surgery." Neuroscientists have pointed out a mismatch between the meager scientific evidence and Canavero's grandiose claims in public appearances, suggesting that he is interested more in drumming up a media circus than in genuinely doing good. Some have called the procedure "egotistical pseudoscience" and "criminal." But if you ask Canavero, he has grand ambitions. Channeling his innermost Dorian Gray, he insists that his pioneering work is a bulwark against the "genocide on a mass scale" that is natural death.

Canavero's first volunteer was a Russian man, Valery Spiridonov. Spiridonov was born with a severe muscle-wasting disorder called Werdnig-Hoffman disease, commonly known as infantile spinal muscular atrophy. The outcomes are usually poor, as the respiratory muscles fail in the first year of life. Spiridonov is one of the few with the disorder who survived past childhood. Spiridonov, now an adult in his thirties, uses a wheelchair for mobility due to the severity of his condition. Optimistically, he was excited and hopeful to benefit from Canavero's experiment. He also recognized that the risks were huge, and he was prepared to donate his body to science. Spiridonov withdrew from the study in 2018, citing his recent marriage and newborn as a reason for living.

* Canavero estimates the total cost of the operation, including the cost of the surgeons, staff, and equipment, to be $100 million.

After the first participant backed out, Canavero's team quickly pivoted. The next closest surgery to demonstrate feasibility would be to swap the bodies of two brain-dead patients, completely unresponsive to their surroundings with no reflexes. The team predicts that with this surgery, they can prove that it is possible to successfully reattach the blood vessels, nerves, and muscles. They also plan to test the patients for head rejection by the immune system.

Likely, the surgery will be performed where medicine follows a different ethical standard than in the West, which has been mostly repulsed by the idea from early on. In the worst-case scenario, we get *The Brain That Wouldn't Die*, a disembodied head kept anesthetized but alive with a steady infusion of oxygenated blood pumped in through carefully connected tubes as the surgical team tries their best to stitch together all the internal connections.

The best-case scenario? We get *Freaky Friday*.

CHAPTER THREE

PEERING THROUGH THE WALL OF SLEEP*

The Secrets of Slumber

STANDING ON ONE SIDE OF THE WALL IS THE WORLD OF THE Awake. Those of us here actively take in all manner of sensory information, from the feeling of clammy air on our skin to sound waves produced by the baying of a faraway wolf to photons of light reflecting off the full moon. We use the combination of our senses and our muscles to purposefully react to stimuli, jumping with a startle when we see a man hideously transform into a werewolf. Fully conscious, we Awake are sensitive and responsive to our world.

On the other side are the Asleep. This is the space where our conscious selves duck into the haze. Our senses are numbed, muffled by the wall of sleep. The strength of incoming information-signaling pathways is dampened, which is why we might sleep through a quiet Monday morning alarm or don't notice if someone next to us is moving while we slumber. To communicate with the Asleep, those external stimuli must be much more intense for the sleeping brain to sense it.

---

* A reference to H. P. Lovecraft's 1919 short story *Beyond the Wall of Sleep*, in which a ruthless murderer is sent to a psychopathic institute where a team of researchers make a distressing cosmic discovery with the help of fancy technology.

But the wall of sleep isn't completely impenetrable. A piece of technology developed in the early 1900s provides scientists and clinicians a tiny glimpse into the sleeping brain. This invention, the electroencephalogram (EEG), is a one-way street, like a telephone on mute. It listens for neuronal activity without sending any electrical currents to the neurons. The EEG uses metallic electrodes placed on the surface of the scalp. These electrodes are so sensitive that they can detect tiny shifts in voltage as small as 10 microvolts—about a million times smaller than the total charge of a nine-volt battery. The tool is sensitive out of necessity since there is a lot of stuff between the electrodes and the outermost brain cell, including hair, scalp, the skull (a quarter of an inch, or six millimeters, thick), and several layers of membranous tissue. Even when applying a salt-gel solution across the electrodes, which improves the conductivity of the electrical signal, only the strongest signals can reach the surface of the head. These barriers mean that electrical changes from single action potentials are far below the detection limit of the EEG. Only the mass activity of several thousand or millions of neurons changing at the same time can be picked up by the electrodes.

A seconds-long track of EEG data looks like random rows of arbitrary squiggles, the trademark activity pattern of neurons chattering away. But after watching brain activity long enough while a person sleeps, clear patterns start to emerge.

At some points during the night, the waves are large, with changes in voltage swinging slowly from tall peaks to deep valleys. To produce these massive deflections, there must be a tremendous number of neurons right at the surface of the brain unified in the timing of their firing activity. Maybe several millions of neurons are all sending action potentials at the same time, causing a massive change in localized electrical potential. This low-frequency activity is called the delta wave. At other times, the brain wave activity looks like that hyperactive chatter of wakefulness: small, sporadic, and jittery. At these moments, the neurons are desynchronized from one another, concerned more with their own activity than with the activity of their neighbors and the network of neurons they are connected to. These two patterns repeat themselves periodically throughout the night at a frequency of about once every 90 minutes. At the beginning of the night, more time is spent in slow delta wave sleep and less time in desynchronized sleep. But, as the morning draws closer, this trend reverses. In a full night's rest of nine hours, a person might spend almost no time in delta sleep during their last sleep cycle.

Looking at other physiological features of the body during sleep along with those brain readings reveals another interesting pattern. Not only does brain wave activity cycle predictably at 90-minute intervals, but so do many other vital statistics. Respiration rate initially dips during the first hour of sleep, falling by 15 percent, and, accordingly, so do blood oxygen levels. Heart rate likewise slows, sometimes by as much as 30 percent. Slowly but surely, these vitals rise over the next 30 minutes until heart rate and breathing begin to resemble the activity of the waking state. This trend repeats every 90 minutes, closely mirroring the patterns of brain wave activity. One of the most obnoxious components of a sleep EEG are the electrodes on the face that measure eye movement. In this aptly named rapid eye movement (REM) sleep, our eyes continue to roam back and forth as if tracking a pendulum swinging erratically. At this point, the large muscle groups of the body are practically paralyzed, although the brain signals look like the person is nearly awake.

The distinction between sleep and wake is a little bit fuzzy at least at the level of the brain. If there is anything separating the two, it's less so a wall and much more a fog.

> This belief had early led me to contemplate the possibility of telepathy or mental communication by means of suitable apparatus, and I had in my college days prepared a set of transmitting and receiving instruments somewhat similar to the cumbrous devices employed in wireless telegraphy at that crude, pre-radio period. —H. P. Lovecraft, *Beyond the Wall of Sleep* (1919)

The EEG has several clinical uses beyond reading brain activity as you explore Dreamlands. If you have had major surgery under anesthesia, the anesthesiologists might have adorned your head with electrodes to monitor your depth of sedation and state of consciousness. When someone is in a coma, physicians use the EEG to constantly monitor the brain, and a change in activity may predict when one may awaken. Epilepsy, brain tumors, stroke, traumatic brain injury—the EEG teaches us something meaningful about all these conditions.

But the man who invented the EEG never would have predicted its many medical applications today. He was too busy trying to prove he was psychic.

A brush with death drove Hans Berger to build the first prototype. Having just dropped out of the University of Jena unsatisfied by his studies in astronomy, he enlisted in the cavalry. During a routine training exercise, Berger's horse unexpectedly bucked, throwing the young soldier in training to the ground. Berger found himself helpless, stunned, right in the path of a massive artillery vehicle rumbling over the terrain. He cried out in panic, fearful for his life no doubt flashing before his eyes. With moments to spare, the operator stopped the vehicle from rolling over Berger.

The next day, much to his surprise, he received an urgent telegram on behalf of his sister, living miles away in his hometown of Coburg. She had wondered how he was doing and asked if he was healthy. A skeptical scientist, adhering to the principles of rigid experimentation and careful observation of events, would note the timing between the two events as a funny coincidence and nothing more. Enlisting in the armed forces carries risks, and horses, no matter how well trained, can still be unpredictable. And it's only natural for family members to connect, especially when a sibling is off doing his military training, a major departure from Berger's low-risk life as a student. There was no reason for any evidence-based scientist to assume that these two events were connected in any way. But a person with a belief in the supernatural might think that they had emitted some spontaneous, telepathic psychic energy during their brief encounter with death. During his time in the cavalry, Berger had never received a telegram from his family. Why now? What was different about this day that made his sister suddenly think of him with enough urgency to send a telegram? Obviously, Berger concluded, his brain generated an invisible distress signal. Luckily for the fields of neuropsychology and medicine, Berger was the second type of person. He believed, on correlational evidence and a questionably small sample size of one, that the brain can emit telepathic signals. It thus became his quest to capture and decode these signals, a quest that culminated in the development of the EEG.

Readings collected from "Berger's machine" were printed mechanically on long strips of light-sensitive paper that were later developed like a photograph. Measuring electrical activity requires a neutral reference point, so the patient would hold a silver spoon in their mouth during the recording. Finally, the clinician would have to pierce the skin and the outer layer of the skull with a metallic probe, a painful process made palatable only under local anesthetic. Over the following years, Berger improved on his design, collecting puncture-free scalp EEG recordings on many others, himself, and his own son in 1924.*

* Modern EEG tech has improved quite a bit in the past century. Simplified EEGs are even available for personal use outside of the clinical setting, allowing anyone with a few hundred dollars to play a variety of "brain reading" games.

After looking through the strips of paper, he noted two different patterns of oscillatory brain activity. The neutral resting signal he called alpha and the higher frequency activity he called beta. The dominant activity pattern changes when a person closes their eyes, performs strenuous mental tasks, or falls asleep. He also noted differences between infants, adults, and older patients; people with a history of seizures; and people with brain tumors. As could be expected, the dead do not show any appreciable brain activity.*

Despite his advancements in functioning imaging technology, Berger was looked down on by his contemporaries in part due to his relative inexperience in the field. His professional background was as a clinician who, his critics argued, had no business encroaching on the academic territory of physiologists. In the end, his naivete became his strength. Physiologists of the time believed that reading such tiny brain signals through the skull was hopeless, and some of them developed complex mathematical formulas just to show how impossible the task would be. Berger ignored the "problems" in the field, paid no attention to the cynics, and ultimately revolutionized medicine.

Tragically, Berger died by suicide in 1941. He had been nominated for the Nobel Prize three times in the year before his death.

LADY MACBETH: Out, damned spot, out, I say! —Shakespeare, *Macbeth*†

* Dr. Adrian Upton arrived at a confusing conclusion in 1974, when he connected an EEG to "America's Favorite Dessert"—Jell-O—and discovered alpha waves indistinguishable from those produced by living, breathing humans. Of course, the intent was to point out the flaws of EEG signal analysis, not to demonstrate the sentience of gelatinous cubes.

† *Macbeth* is full of witch curses and ghosts. Even speaking the name "Macbeth" inside a theater is an invitation to disaster, hence the use of "The Scottish Play." These superstitious thespians might be onto something, as productions of *Macbeth* have been plagued with misfortunes: a theater manager dying right before opening night, an onstage murder when a prop knife was switched for a real blade, and a riot between rival performers that led to the deployment of militia who accidentally shot and killed 31 theatergoers.

The guilt-ridden Lady Macbeth, frantically washing her hands while asleep, is just one fictionalized account of a person who gently walks that line between wake and sleep. Sometimes, that walking is not gentle at all.

For most people, sleep is characterized by a general dip in physical activity. But the body isn't entirely paralyzed for the entire night. It's totally normal for someone to move around a little, even in response to some input stimuli. Is a cold night breeze invading your bedroom? You may pull yourself under the warmth of your covers. Maybe your body has been in the same position for too long. Rolling over can help shift your muscles. These examples of motor activity are subtle, expected, and harmless.

But some people experience somnambulism, or sleepwalking, where their night's rest may be interrupted by robust physical activity. With a lifetime prevalence of about 7 percent, somnambulism seems to be more common among children—think of the boy Dylan, haunted by Freddy Kreuger in *Wes Craven's New Nightmare* as he sleepwalks across a busy highway. Many outgrow their nocturnal wandering sessions, however: by adulthood, only 1.5 percent of people continue to sleepwalk. And while it's not known exactly what causes it, there are some hints that illness, sleep deprivation, or extreme tiredness may trigger an event. Additionally, there is some influence of genetics since identical twins are more likely to sleepwalk compared to fraternal twins or siblings.

The actions of sleepwalkers range in complexity. Innocuous simple activities include raising the arms, opening the eyes, sitting up in bed, or turning to rummage through a nearby nightstand. Somniloquy, or sleep talking, is not uncommon.* Other sleepwalkers may coordinate many body systems, resulting in highly intricate, complex activities. Sleepwalkers may make their way down

---

* I am prone to doing this, often to the distress, confusion, but mostly amusement of anyone around me. Allegedly, I once had a full sleep conversation between multiple people where I acted out all the people using different voices. Allegedly.

a flight of stairs, navigate around obstacles, open the refrigerator, and prepare themselves a sandwich as a late-night snack, complete with a side of sliced fruit. They may do completely unexpected things, such as rearranging furniture. Medieval records of sleepwalkers even describe nobles who rose from their beds, mounted their horses, took to the grassy plains, and swung their swords at some imaginary foe. And then there is a diagnosis that requires no explanation: sexsomnia.

In the case of Lee Hadwin, his sleeping self is a talented artist, much in contrast to his waking self, who claims to have no artistic talent and the poor grades to back it up. His sleep drawing began in childhood, when he scribbled all over the walls and floors in the middle of the night, much to the confusion of his parents. At 15, he awoke to find a beautiful, lifelike portrait of Marilyn Monroe, who has since become the muse for his artistry. He says that the sleep artist comes out a couple times a week. Like any other artist, Hadwin has even suffered "artist's block" for two years, when his unconscious dozing self produced nothing.

Even though most nighttime roaming is harmless, a handful of case studies demonstrate that sleepwalkers might be a danger to themselves or others. Consider the following:

- A three-year-old girl with nocturnal trichotillomania, the compulsive pulling of her own hair, combined with trichophagia, the eating of that hair.
- A 37-year-old man who, in a demonstration that the statistician playing darts is wrong,* not only ate large portions of raw meat one night, leading to unpleasant tummy issues in the morning, but also on one occasion burned the food he was cooking on the stove.

* The first dart goes to the left. The second dart goes to the right. The statistician exclaims, "Bull's-eye!"

- A 45-year-old woman who developed a habit of sleep smoking. She would wake in the morning to discover burn holes in her bedsheets and blisters on her foot from stepping on lit cigarette butts.

And there's the very surprising case of Kenneth James Parks, who went to sleep on May 23, 1987, and woke up the next day a murderer. At some point in the middle of the night, Parks got into his vehicle. He drove 14 miles (about 23 kilometers) southwest to the town of Scarborough, outside of Ontario, Canada, to the house of his in-laws. Brandishing a tire iron, he opened the door with his key, made his way to the bedroom, and viciously attacked the sleeping couple. Mrs. Woods died from blunt-force trauma and knife wounds. Mr. Woods survived but was severely injured. Then Parks drove himself to a police station where he turns himself in. Covered in blood spatter, he announced in a panic, "I think I have killed some people."

Parks was charged with one count of murder and one count of attempted murder. Leading up to the trial, every aspect of his life was scrutinized. He had a good relationship with the victims. His wife testified that they were supportive of Kenneth. He had incurred tremendous gambling debts, leading him to put his house up for sale. To cover a shortfall, he stole money from his work. When he was caught, he was subsequently fired. The embezzlement trial was coming up, and he was planning to tell his in-laws about his gambling addiction, both certainly sources of stress that led to a new onset of insomnia. Of course, there were no EEG data from the night when the crimes were committed. But recordings collected later discovered unusual patterns of activity, including several wakings each night (zero to four in a healthy sleeper, up to 10 in Parks) and far more shifts between sleep stages (25 to 60 in healthy people, 170 in Parks). The jury weighed this EEG evidence, and he was acquitted on both charges.*

There are a handful of other cases of homicidal sleepwalkers, some of whom were also found not guilty. The first successful use of this defense in the United States was in 1846 by a Bostonian named Albert Tirrell. Tirrell left a horrendous crime scene. The victim, his mistress Mary Ann Bickford, was discovered with a massive gash across her neck, her clothes and hair singed black. He fled the city, bouncing between Canada, New York, and finally New Orleans, where he was apprehended by the authorities. At his murder trial, family members testified of Tirrell's long track record of sleepwalking. A cousin recounted a bizarre incident when Tirrell had pulled him out of bed and threatened him at knifepoint, an event that Tirrell had no recollection of. The dean of the Harvard Medical School testified that sleep killing could, in theory, be possible. The jurors agreed that there was not sufficient evidence to convict him of murder.

In 1925 in Texas, Isom Bradley was charged with the murder of his girlfriend. In the middle of the night, Bradley had grabbed his gun and shot someone whom he believed had broken into the house. When he regained awareness, he was holding a gun, and she was lying dead at the foot of the bed. His defense argued that somnambulists lack the "free and rational exercise of [their] understandings." Therefore, because he was not in control of his conscious power of volition, he should not be held morally responsible for his actions. The jury agreed.

Much rarer than common sleepwalking, patients with a diagnosis of REM sleep behavior disorder (RBD) present with similar symptoms. It may be more dangerous than ordinary sleepwalking

* If this story sounds familiar, Parks was the inspiration for *The Sleepwalker Killing*, a 1997 film featuring Hilary Swank.

for two main reasons. In RBD, people often have no awareness of their surroundings, unlike the sleepwalkers who can safely navigate their surroundings. Second, RBD takes hold during REM sleep, the phase when a person is likely to dream. These two factors combined make for a dangerous combination since injuries are more likely when someone acts out their exciting slasher-flick nightmares filled with harrowing action sequences. Consider the following cases:

- A man, presumably the hero of a summer blockbuster, swung his arms wildly, accidentally punching his sleeping spouse. In some cases, sleep fighting resulted in spouses needing dental repairs or, in a weird twist of irony, spouses being bitten.
- Acting out a football dream, a 77-year-old patient caught a punt at the cost of face-planting off the bed. The subdural hematoma that formed following this trauma was treated by drilling a hole in his skull.
- In dream-defending themself from a dream-bear attack, one patient threw their daughter across the room. Another patient kicked a hole in the bedroom wall. Yet another discharged a firearm.
- A woman dreamed of being chased by an out-of-control steamroller. To avoid being crushed, she swan dived off a five-foot-high platform (1.6 meters), resulting in a brain bleed.

Harmless sleep talking, homicidal sleep killing, and RBD dream acting are all ways in which a person straddles the wall of sleep. But are there any scientific explanations for why some people wander around in the late hours of the night when all the good citizens should be asleep?

Early explanations were proposed in *Malleus Maleficarum*, a classic fifteenth-century treatise on witchcraft. The specific witch power* that is most blamed is possession of a sleeping body. Without the victim's awareness, some unholy devil would lift the bewitched person through the sky, maybe dancing their bodies across rooftops. Once someone spoke the baptismal name of the possessed, the curse would break immediately, causing the victim to fall.† People who sometimes awoke in the middle of the night standing in waist-deep water with no recollection of how or why they got there were more likely somnambulists.

This theory, however, offers little to the rational-minded folk of the Enlightenment and beyond. Sleep labs first started appearing in the 1970s as medicine began to acknowledge the profound impact of sleep on health. These labs swapped out pipettes and cold benchtops for an environment more conducive to sleep: a bed, a nightstand, and maybe a decorative house plant or frilly pillow, akin to a comfortable bedroom where the sleepwalkers can feel at home. Modern sleep labs are also equipped with a whole host of recording equipment, including video cameras, respiration and blood oxygen level monitors, and a cap studded with EEG electrodes for measuring changes in brain activity. These sleep labs have revealed some clues about what might predict sleepwalking, including accumulated sleep debt, migraine disorder, fever, or even those racked with guilt of any severity, ranging from shoplifting to Shakespearean conspiracy to murder. Despite all the technological advances in biometric measurements and brain imaging, no definitive, neuronal-level explanation can account for somnambulism.

But a theoretical framework has been proposed based on observations of case studies of individual sleepwalkers and some

* This book also contains a wonderful section that blames witches for a man's inability to maintain an erection. Obviously, the book was written by a man.

† Mina Harker from Bram Stoker's *Dracula* specifically mentions this belief in a letter dated July 26. There is quite a lot of sleepwalking throughout the tale, including what could be the first suggestion that the trait runs in families.

nonhuman experiments. At the heart of this framework is the principle that different brain structures contribute to a specific aspect of normal sleep activity. During a healthy night's rest, these circuits turn on or off at a very specific pattern with precise timing, like a mosaic of programmed neural activity. But in people with parasomnias, the individual parts do not contribute to their relevant circuits when they are supposed to, leading to disorder of the normal sleep machinery.

As an example, consider this model in sleepwalking. You expect that both consciousness and voluntary motor circuits are tethered to one another. As we drift off, we lose awareness of our surroundings and decrease movement, as if consciousness and motor circuits are two lightbulbs controlled by the same switch. Deactivation of motor circuits is secondary, achieved by turning up an inhibitory system that originates in the brain stem. These motor inhibition systems are most active during REM sleep, likely with the intent of preventing people from acting out their dreams. When a person sleepwalks, sleep draws, or sleep fights, those activity circuits are not shut down at the same time as the consciousness circuits. The switch is turned off, and consciousness follows suit, but the motor activity light stays on.

This theory is supported by some experimental evidence. One part of the brain stem, the dorsal pons, was zapped out in cats. Tiny cat EEGs were set up to record their brain activity. The cats would doze off to sleep, and their brains would go into REM activity, but their muscles wouldn't lose tone as you'd expect. Instead, they moved around as if they were awake. Some of the cats would stretch out their paws and rock their heads back and forth. In other cats, they would fully raise their heads and forearms in a searching and scanning pattern, as if the dreaming cat were tracking a fly in their slumber. The most interesting of these cats pounced and clawed at the floor, acting out an exciting dream of stalking and hunting a mouse (presumably).* Our analogous brain stem circuits are the motor circuit inhibitors, the ones that prevent us from punching, biting, or killing our loved ones while we sleep.

Prescription insomnia drugs such as benzodiazepines increase the likelihood of sleepwalking. These drugs affect our brain's neurochemistry by changing the activity of the neurotransmitter GABA, which is the main inhibitory signaling molecule in the brain. When GABA binds to receptors, it makes it more difficult for neurons to fire an action potential. At the molecular level, benzodiazepines make the GABA signal more powerful. Normally, GABA shushes other neurons using an indoor voice; in the presence of a benzodiazepine, GABA screams. By enhancing the inhibitory effect of GABA, these hypnotics calm brain activity, lulling you into slumber.† But they also distort the sleep mosaic in weird ways. One subset of sleep aids, particularly the "z-drugs," has gained a notorious reputation for inciting sleepwalking. This side effect is so well documented that the U.S. Food and Drug Administration added a boxed warning informing patients of their elevated risk of these complex sleep behaviors. Boxed warnings often have a backstory, and the list of reports that led to this box include a handful of fatal motor vehicle accidents, falls from buildings, a 14-year-old-boy who spilled gasoline all over the garage while trying to fill a lawnmower, and a man who shot himself in the head with a .22-caliber handgun.

What happens when this mosaic is flipped and the consciousness circuits turn on while voluntary motor control stays off? This experience is sleep paralysis. About one-fifth of people have felt sleep paralysis at least once in their life: a temporary state when a person wakes but is unable to move their body. For most

* Although according to Neil Gaiman's "A Dream of a Thousand Cats" from *The Sandman*, this cat may have stalked, toyed with, and killed a very different mammal.

† These drugs also help people cope with disorders that have their origins in nervous system hyperexcitability: epilepsy, anxiety, muscle spasms, or alcohol withdrawal.

people, this feeling of helplessness is intensely terrifying. Sometimes, imagery from dreams bleeds into waking existence, which explains why so many cultures project their fears and superstitions onto the sensation. Some link alien abduction reports with sleep paralysis. Egyptians attributed sleep paralysis to bodily possession by jinni, powerful spiritual creatures from Islamic folklore. Some Italians have blamed a humanoid cat or a ghost witch as the cause of their temporary paralysis. Some Cambodian refugees, a population highly vulnerable to sleep paralysis, report seeing a fanged demon pushing them down or a dead spouse wrapping their hands around the throat of the victim.

Figure 3.1. Difficulty with breathing and visual hallucinations, common symptoms during a bout of sleep paralysis, are monster-ified in Henry Fuseli's *The Nightmare*, taking the shape of a demon sitting on the sleeping woman's chest. TULIP HYSTERIA, CC BY 2.5 HTTPS://CREATIVECOMMONS.ORG/LICENSES/BY/2.5, VIA WIKIMEDIA COMMONS

A third piece of the sleep mosaic are those circuits that drive involuntary muscles, such as the nerves for respiration. Many of the earliest sleep labs were built for diagnosing sleep apnea, a common condition where people are unable to breathe in the middle of the night, causing them to wake and leaving them tired in the morning. While most cases of sleep apnea result from a physical blockage of the upper airway, a handful of cases are a result of faulty brain circuitry. The involuntary respiration circuits of the brain stem are supposed to be on autopilot, firing all the time independently of whether the sleep switch is on or off. In central sleep apnea, those respiratory circuits become tethered to certain phases of sleep, deactivating when they're not supposed to. The more extreme version, congenital central hypoventilation syndrome, is fatal without nightly medical interventions. The condition is also called Ondyne's curse: a reference to the mythical water nymph who cursed her cheating husband, who pledged "every waking breath" to her. Some miswiring in these circuits, some misplaced tiles in the mosaic, and we forget to breathe.

And then there are the people whose sleep switch flips up when it's not supposed to; 0.1 percent of people have narcolepsy, characterized by excessive daytime sleepiness, leading to their falling asleep at inopportune times. In severe cases, narcolepsy may be accompanied by cataplexy, a sudden muscle weakness that causes a person to lose all muscle tone and slump to the floor. One of the most common triggers for sleep attacks are high emotional states, including laughter and stress. A neurochemical signal called orexin* is a major contributor to these uncontrollable sleep attacks. In comparing the brains of sleep-normal and

* In 1998, two research labs discovered the same chemical signal. Confusingly, each of them gave the hormone a different name. One lab called the substance orexin, named because of the cqntent in which they described it: hunger. Increasing orexin levels caused an increase in feeding behavior. The other lab called it hypocretin, a portmanteau of its brain structure of origin, the *hypo*thalamus, and the name of another hormone that was already known at the time, se*cretin*.

narcoleptic people, the narcoleptic brains had less than one-tenth of the total orexin neurons than the healthy patients did, representing a dramatic loss of cells. Orexin may be the glue that holds the rest of the mosaic in place. But considering how intricate the whole system is and how many things can possibly go wrong as we cross the "wall of sleep," it's evident there is much more than meets the rapidly moving eye.

> In the moonlight opposite me were three young women, ladies by their dress and manner. I thought at the time that I must be dreaming when I saw them, for, though the moonlight was behind them, they threw no shadow on the floor. —Bram Stoker, *Dracula*

Dr. Sigmund Schlomo Freud (1856–1939) simultaneously holds two titles in modern psychology: most famous and most wrong. As for his accomplishments, the bearded, cigar-smoking neurologist was a major player in the development of psychology as an independent field of study, sort of bridging the gap between the "cells and chemicals" approach espoused by biologists and the purely abstract realm of the mentalists. He advocated that the brain was the root of the mind, a paradigm shift that eventually led to our modern field of neuroscience. His most enduring legacy was a new approach to therapy called psychoanalysis, a method that focuses on dissecting out subconscious thoughts to determine the root of a patient's neurological distress. The emphasis on the influence of early childhood trauma on adult behavior is still a major point of discussion in modern psychology. Today's forms of talk therapy, including cognitive behavioral therapy, may have taken some of their cues from Freudian psychoanalysis.

Despite his contributions to our understanding of the mind, his appearance in every high school psychology textbook, and his multiple additions to the American lexicon,* the man was wrong about nearly everything. His Oedipus complex theory, which suggests that a young man's first sexual desire is his mother and that this erotic drive motivates a jealous hatred of his father, is difficult to take seriously. He believed that a woman's societal role was entirely dependent on her reproductive capability and that women envy the penis and blame their mothers for not giving them the best possible "tools" for life.† Psychoanalysis was especially criticized since it lacked objective, quantifiable evidence. Without a rigorous application of the practice, a psychoanalyst had tremendous flexibility when making a diagnosis, essentially steering the session toward any outcome.

His most groundbreaking work, the massive *Die Traumdeutung* (*The Interpretation of Dreams*), is now regarded as bad science based on the last of the criticisms. Within the pages, Freud lays the foundation for his famous theory of the psyche. According to Freud, every mind has three agents that influence the actions of the individual: id, ego, and superego. The id represents our unconscious, animalistic nature, acting strictly to fulfill basic needs and desires while maximizing pleasure and minimizing pain. The superego is the opposite, operating consciously in conjunction with learned societal norms and parental upbringing. The ego has the unenviable task of resolving the conflict between the two by considering the demands of the id and subsequently mediating behavior to act in a socially acceptable manner. According to this construct, the ego constantly represses the id.

---

* Freudian slip, an accident made while speaking when a person says something that reveals a subconscious desire, generally as it relates to sex; phallic symbolism, the idea that long objects are representative of the male sex organ; and libido, the drive that encourages a person to pursue sex, are all psychological concepts invented or popularized by Freud. The man had a one-track mind.

† Freud completely misplaced the blame. It wasn't until a 1959 publication after Freud's death that geneticists developed the theory that the Y chromosome contributes to the development of sex organs. The Y chromosome is passed down by the father, not the mother.

This conflict is the cornerstone of Freud's theories about all human behavior, including dream content. Throughout the night, often during REM sleep, we experience vivid, multimodal hallucinations. They are typically disjointed and nonsensical, rarely telling a cohesive story. Some dreams are mundane and lacking in emotional content, maybe a replay of the previous day's trip to the grocery store or a parade of the faces you haven't seen in years. Some dreams are more dynamic, provoking the whole range of genuine emotional states on waking, ranging from absolute panic to all-consuming infatuation. They can elicit physiological changes, causing you to wake in a sweat over a college exam you forgot to study for or with your heart pounding after running away from a mob of fish-like cultists.

Within *Interpretation*, Freud provides a framework for deconstructing the content that the brain produces while we dream. As a clinical psychiatrist in the 1880s, Freud had worked with patients with various psychoses. He used their pathological hallucinations to try to explain dreams more broadly. To understand a person's desires, Freudian dream analysis pays special attention to two aspects of the dream hallucination that he called the manifest content and the latent content. Manifest content is what we recall from our dreams on waking. The manifest content is often strange. Freud suggested that free association is one useful tool to get at latent content: the hidden information that the dreamer's subconscious is expressing. In talking out experiences without any specific end goal, the patient reveals their unconscious desires.

According to Freud, dreams are a disguised representation of one's wishes, distorted due to "a kind of inner dishonesty." So when someone such as *Dracula* protagonist Jonathan Harker experiences a raunchy dream, Freud would attribute it to his id. This part of Harker expresses a desire for sexual contact, especially as he is on a trip away from his beloved fiancée, Mina. But the ego is constantly keeping him from acting out that desire. However, once he passes through the wall of sleep, Harker's id is now free from that repression. What we see in his dream is therefore a Freudian expression of his temptations made real, causing him to feel "a wicked, burning desire that [the three unknown ladies] would kiss me with those red lips." Unbridled by the constraints of Victorian societal norms, Harker's id is wild and free during his dreams. Were Lieutenant Ellen Ripley's recurring nightmares of being the host to a xenomorph chest burster an expression of her desire to get pregnant after finding out her daughter had died while she was in cryostasis? When Sue, Carrie's remorseful and reformed bully, is grabbed from beneath a pile of rubble in the closing dream sequence, is it because she wishes that Carrie would accept her apologies and embrace as friends? Were Freud alive to see *Aliens* or *Carrie*, he might say yes.

At the end of all the nonsensical free associating and phallic object identification of Freudian dream analysis, H. P. Lovecraft's words from *Hypnos* resound: "Wise men have interpreted dreams, and the gods have laughed." Sometimes, a wooden stake is just a wooden stake.

> People think dreams aren't real just because they aren't made of matter, of particles. Dreams are real. But they are made of viewpoints, of images, of memories and puns and lost hopes.
> —Neil Gaiman, *The Sandman*

According to the Freudian dream interpretation framework, people dream to resolve repressed desires, working through their issues by acting through them in a protected space free of consequences. But do scientists in the post-psychoanalysis era have any explanations for why dreams happen?

Short answer: no.

Long answer: whatever evidence is out there has not produced a strong consensus. A prominent theory suggests that dreams help us reinforce both spatial memories, such as navigating a new environment, and implicit memories, which are "muscle memory" skills, such as riding a bike. This hypothesis proposes that patterns of neural activity reactivate during REM sleep like an echo, allowing us to rehearse passing through those surroundings or practicing those skills while we sleep. But this hypothesis tells only part of the story. Dreams often have some spatial content, but those weird dream high school hallways and classrooms are usually wildly discontinuous and hardly ever accurately mirror any environments we have experienced, making it ineffective as a rehearsal tool. And there are all sorts of conditions where REM sleep is disrupted or abolished, such as when taking antidepressants or following brain injury. In these cases, there is less active dream recall, but people continue to live totally normal and healthy lives without a noticeable decrease in their ability to learn. As an additional criticism of the learning theory, we practically never dream about the stuff we end up remembering. I frequently dream about not knowing where my classroom is or that my PhD can get revoked if I lose at brain trivia. I can't think of a single dream where I was juggling.*

Diametrically opposed to this theory is the proposition that dream sleep is when *un*learning takes place. Every day, we encounter millions of stimuli, from the earthy acrid tones of our morning coffee to feeling the cool side of the pillow on your cheek right before drifting off. Most of that stuff doesn't get any attention; practically none of it is of any consequence tomorrow. Looking even further backward, there are probably older memories that go unforgotten. Maybe the brain has randomly decided that it has become too much of an energetic upkeep cost to remember the face of a friend from elementary school or the floor plan of a mall you last visited decades ago. At the level of neurons, this forgetting

* Granted, we often recall only a tiny tidbit of our dreams. It's possible that the other 95 percent of my dreams were circus clown dreams.

might happen when dendritic spines shrink or even disappear. Perhaps these circuits get activated briefly right before their destruction, causing us to hallucinate running into Bethany at the Fayette Mall. This unlearning theory of dreams, published in *Nature*, one of the most exclusive scientific journals, gained traction partly because of who proposed the idea: Nobel Prize winner Francis Crick of DNA-is-shaped-like-a-double-helix fame.

While on the topic of neural plasticity, a relatively young theory suggests that dreaming carries out a slightly different function. The neurons of the cortex, the outer surface of the brain, exhibit a remarkable capacity for change. This is most evident among people who have lost their vision. For sighted people, the information collected from our eyes gets processed by the visual cortex, a series of structures in the occipital lobe. But when a person goes blind at the level of the eyes, maybe because of an accident or disease, these brain areas do not go completely to waste. Instead, other sense organs start using these circuits. Blind people can learn to echolocate by making clicking sounds, using the reflection of those clicks to create a mental map of their surroundings.* Ask these people to click while in a brain scanner, and their visual cortex shows patterns of activity just as if they were seeing. This plasticity also happens when blind people read Braille: Their touch circuits get rewired into the visual cortex. The cortex is precious real estate; nothing gets wasted. In both examples, plasticity is a good thing since the visual cortex helps with human echolocation or during comprehension of tactile language. But you might not always want plasticity. What if during sleep, when our visual inputs are negligible, our visual cortex is subject to takeover by our other senses? One fascinating proposition suggests that dreams, which robustly recruit and activate visual circuits, serve as a bulwark against the steady advancing creep of other circuits.

* As Lucius Fox says to Bruce in *The Dark Knight*, "Like a submarine, Mr. Wayne. Like a submarine."

Those wildly imaginative dreams populated with talking silverware or flying cat–horse hybrids might serve a biological purpose, more so than to provide interesting conversation topics. Some suggest that these hallucinations help us flex our creativity, something that so often falls into disuse during the toxic mundanity of adulthood. When we specialize, we get very good at a handful of things, but we forget how to apply that knowledge to other domains. Dreams enhance our creativity, preventing our brains from becoming overfitted. From an evolutionary perspective, our ancestors were able to rapidly adapt to new situations, and mental flexibility remains one of the biggest strengths of our brains.

The reality behind this research is that dreams are slippery and ephemeral, making them notoriously difficult to study. Dreams are one of the few windows into the details of the sleeping brain. The other windows we have, such as brain-imaging scans (e.g., EEG or fMRI), are unable to see single cells let alone the tiny spines along those cells. We could get this level of resolution experimentally in monkeys, but since dreaming is a subjective experience, studying their sleeping brains doesn't offer much insight. In addition, there's the issue that perception of time in dreams is profoundly distorted. What feels like a three-act narrative complete with fully fleshed-out characters, a specific setting in place and time, and internal motivations may whiz through our brains in minutes or seconds. Self-report while awake is already notoriously unreliable and significantly more so when we sleep.

There is a lot of fuzziness behind the science of dreams. Whatever is happening in our brains when we dream is driven by something real, something tangible. Maybe dreams exist for something necessary and biological, and maybe they reveal something about our subconscious; both are probably true. In the end, no one can deny that dreams captivate our imaginations and leave us feeling our most genuine human emotions when we wake. Without the creativity brought on by dreams, our world would be a little bit emptier. Try to guess what the following dreams inspired, arranged from easiest to most obscure:

- ". . . the hideous phantasm of a man stretched out . . . how signs of life, and stir with an uneasy, half vital motion."
- ". . . metal death figure coming out of a fire."
- "No, I named her in the spirit of fan love, which is the purest love there is. You should be flattered."
- ". . . whirl me through space at a sickening rate of speed, the while fretting & impelling me with their detestable tridents."
- "The mind's eye . . . now distinguished larger shapes of varied shapes, many times closer together; everything in motion, twisting and turning like a snake. . . . The figure whirled mockingly before my eyes" (this reference is unlike the others, so don't feel disheartened if nothing comes to mind).*

So Alice got up and ran off, thinking while she ran, as well she might, what a wonderful dream it had been. —Lewis Carroll, *Alice in Wonderland*

---

* The inspirations and the dreamers are, in order, *Frankenstein*, Mary Shelley; *Terminator*, James Cameron; the main antagonist in *Misery*, as it came to Stephen King on an airplane dream; Night-Gaunts, the beings described in *The Dream-Quest of Unknown Kadath* that were based on the many nightmares of a young H. P. Lovecraft; and the chemical structure of the benzene ring, a huge challenge in the field of organic chemistry, as inspired by August Kekulé's dream of an ouroboros.

CHAPTER FOUR

"HAPPINESS IS KNOWING YOU GOT SOME PILLS"*

FOOD, DRUGS, AND OTHER THINGS THAT MAKE US HAPPY

HAPPINESS IS A TRICKY CONCEPT. SHORT-TERM AND LONG-TERM happiness look like different things. Eating a third or fourth donut gives that burst of "oh yeeeaaah," but so does seeing fitness goals manifest over the course of months. Both concrete and abstract things can make us happy. Finding a wadded-up $10 bill in your pocket surely puts a smile on your face, but so does finding out that your investment portfolio just went up 15 percent. Usually, it's good things that we are happy about. But in an excellent example of "there's a German word for that," we have *Schadenfreude*, deriving happiness when bad things happen to other people. All this is to say that happiness is complex, and that the emergent properties of what it means to be happy are not easily explained on the cellular level no matter how much we know about action potentials and neurotransmitters.

There is one aspect of happiness, however, that neuroscientists feel comfortable enough to quantify, and that is pleasure. Just to be

* Philip K. Dick's *A Scanner Darkly* (1977) plays out as a firsthand account of the consequences of drug misuse. His drug of choice was amphetamine, or speed. He took up to 1,000 pills a week and claimed to have written 68 pages of prose in a day.

clear, no one is claiming that happiness and pleasure are the same thing. Pleasure can lead to happiness, but not all happiness comes from pleasure. Typically, we think about pleasure as being short term with a strong sensory component—that second, third, or fourth slice of chocolate cake is a great example of what is pleasurable. Studying pleasure in a laboratory turns out to be a relatively straightforward task, and it starts with watching animals in a box.

It was a pleasure to burn. —Ray Bradbury, *Fahrenheit 451*

A definition to clarify before talking about pleasure is the difference between a primary and a secondary reinforcer. Primary reinforcers are things that provide tangible, intrinsic value to the animal, whether for survival or for pleasure. Some examples include water, food, sex, and social contact. Humans and rats are not too different with respect to primary reinforcers since both crave these things. A secondary reinforcer is something that an animal learns to associate with reward, sometimes neutral things that may look nothing like the primary reinforcers. The clearest example of a secondary reinforcer in human society is money, which has no inherent survival value but still brings people tremendous pleasure. Money represents the ability to acquire survival needs and wants, but you cannot survive on money alone. A box full of Arrakeen melange is worth very little if your stillsuit is ripped.*

Abstract secondary reinforcers can be learned and trained through societal cues and repeat experiences. Take grades, for example. Early in life, many of us are taught that "high grades = good." Your hunger doesn't go away after getting an A, and a B+ won't help

* In Frank Herbert's *Dune*, the spice melange is the most precious resource in the universe. However, it comes only from the harsh desert planet of Arrakis, where the natives wear stillsuits to preserve the body's precious moisture.

dehydration any more than a B–. But maybe we get rewarded with praise for bringing home an A, which encourages future high numbers. The excitement of acing a math quiz and the dejection from an 89.4 percent are trained experiences where we associate a secondary reinforcer with some kind of emotional response.

B. F. Skinner, the "Father of Behaviorism," demonstrated in the laboratory that nonhumans too can also learn to associate neutral secondary reinforcers with a primary reward. The apparatus that bears his name, the Skinner box, is a rat cage that contains a lever. To the rats, this is the first time they have ever come across this lever. So, being the naturally curious animals they are, they walk over and begin to engage with the lever, maybe by putting their paws on top and pushing down or even by accident. When they do, a little morsel of food drops into a receptacle in their cage. Rats very quickly learn that their survival might depend on that lever. This type of task where the animals learn to actively perform some task to get some outcome is called operant conditioning as opposed to Ivan Pavlov's passive conditioning. Skinner wanted to show that associating a reward with a secondary reinforcer can promote the learning of completely novel behaviors, so he taught his pigeons to play competitive table tennis. Two pigeons facing each other will peck at a ping-pong ball on a gently sloped table, back and forth, until one pigeon fails to properly predict the trajectory of the ball. Once it rolls off the edge, a little pellet of bird food is awarded to the victor.

Skinner's next undertaking was training pigeons to be, naturally, suicide bombers. The theory behind the government-funded Project Pigeon was straightforward. First, train some pigeons to peck on a target of interest displayed on a screen. Then put those pigeons in the nose cone of a glide bomb. Connect a screen to the course-correcting steering system in case the payload veers off course and let the pigeons correct for variables like wind speed. Not quite the precision drones of the 2000s, but for the 1940s, it was a reasonable attempt at distance warfare, something that would keep American soldiers out of harm's way while simultaneously attacking Axis aggressors. The project may have had a future had the higher-ups taken the design more seriously, but it was an uphill battle to convince the brass that the only thing standing between a successful strike and unnecessary civilian deaths was a couple of birdbrains. Not to mention, funding was limited, and pigeon-guided bombs had to compete with the overwhelming destructive power of the Manhattan Project, which promised much more dramatic results. In addition, successful harnessing of nuclear technology had civilian applications outside of warfare. Pigeon power, however, didn't.

Skinner's most grandiose claim from his research is summarized in his ominously titled *Beyond Freedom and Dignity*. Here, he argues that animals, from lab rats and pigeons all the way up to humans, ultimately have no free will. He insists that everything we do is a result of a lifetime of learned associations between some secondary reinforcers and outcomes. Every stimulus in our environment provokes either an approach or an avoidance so that we maximize our access to primary reinforcers.*

Skinner boxes have undergone a few improvements over the years. Some boxes are equipped to pair the lever with a unique light and sound tone that indicate when the lever becomes active. The addition of classical Pavlovian cues helps the animal learn the behavior faster. The types of questions tested by the experiments also changed. Normally, lab rats and mice are given unlimited access to a standard diet of chow. Although chow is an ideal blend of all the nutrients and calories they need, the food is compressed into light beige pellets that smell like a barn and aren't particularly pleasant to the human palate.† But put the animals on a 24-hour

* At times, *Beyond Freedom and Dignity* feels a bit radical and diminutive of humanity's capacity for flexibility. At other times, it feels like an advertisement for *Walden Two*, a work of utopian science fiction that he published prior.

† Pardon this editorialization.

fast before Skinner box testing? They lever press frantically, unsure if they will lose access to their food source again. Inject them with leptin, the satiety hormone that is produced when we are full? They lever press less often even if they haven't eaten in a while, showing that this chemical signal is powerful enough to override their homeostatic need for calories and nutrition. Give them access to pellets with six times the normal fat/sugar content? Aggressive lever pressing is followed by gorging themselves silly.

Food is one of the most basic and simplest of the primary reinforcers. If you swap this out for other rewards, you can observe how much an animal desires or how much they are willing to work for various stimuli. Mice lever press to gain access to other mice, suggesting that they crave and are willing to work for social interaction. In a paradigm where the floor is equipped to deliver a little electric shock and lever pressing prevents that shock, they will learn to press to avoid pain.*

> It's like life—it presents a different face each time you take it. Some hold that the spice produces a learned-flavor reaction. The body, learning a thing is good for it, interprets the flavor as pleasurable—slightly euphoric. —Frank Herbert, *Dune*

As you might predict, the Skinner box is a tool for behavioral neuroscientists to study recreational drugs, such as the spice melange from *Dune*, Substance D from *A Scanner Darkly*, soma from *Brave New World*, genre from *Westworld*, or one of any of the hundreds of chemicals considered by drug regulatory agencies to be controlled substances. Modifications of behavioral tests where nonhumans learn to give themselves drugs is called self-administration.

* In support of Skinner's philosophy of radical behaviorism, humans also can learn to take pleasure in engaging with secondary reinforcers, perhaps by dousing houses in kerosene and then lighting them afire.

With a steady hand and a little bit of practice, a skilled (rat) surgeon can thread a flexible plastic catheter directly through the jugular vein. The other end of the tube is connected to a syringe filled with a drug controlled by an automated pump. After recovery, the animal is put into a Skinner box, where pressing the lever causes a little infusion of the drug straight into the bloodstream. Rats quickly learn to lever press for a dose of many drugs that humans take recreationally, ranging from cocaine and heroin to nicotine and ethanol.

Early self-administration studies asked the question "At what point do monkeys stop self-administering drugs?" The final end point, for some drugs, is death. In one monkey experiment, animals were trained to deliver an intravenous injection of a variety of drugs, such as morphine, amphetamine, phenobarbital, caffeine, and three drugs that resulted in self-administration until death: cocaine, alcohol, and codeine, an opioid commonly found in cough syrup. These controversial experiments would not be able to be done today: They were published in 1969, one year before the development of government-mandated regulations on animal welfare in research and the establishment of modern ethical guidelines for such work.

Beyond the question of toxicity and limits, self-administration is also used for studying the variables that affect drug use and misuse with the goal of developing treatments for people. As always, translating behavioral observations from nonhumans to people is the biggest challenge in the field, and this is especially true in studying substance use disorder. Human drug use is a much more complicated phenomenon than can ever be modeled in lab rats with their simple, static surroundings. Clinically, humans are diagnosed with substance use disorder using a series of criteria that fall broadly into four categories. With a little creativity, modifications to the self-administration paradigm can allow for loose approximations of these four categories:

1. A person has a difficult time controlling their use of the substance. This looks like a person who uses more or for a longer time than they intend on. Maybe they regularly snowball off one beer into an entire night and morning. They may also relapse from abstinence frequently, an experience best summed up by Mark Twain: "Giving up smoking is the easiest thing in the world. I know because I've done it thousands of times." Perhaps they invest a great deal of time and energy to obtain the drug, such as those who would drive hundreds of miles to visit a morally corrupt physician who writes prescriptions for opioid painkillers promiscuously with little due diligence. To model this "whatever it takes" attitude, rats in Skinner boxes receive drugs on a progressive ratio schedule, meaning that it requires exponentially more lever presses to get the rewards: whereas the first lever press would deliver the first dose, 12 presses would be necessary for the fifth dose and more than a hundred presses for the fifteenth. Relapse can be modeled using an extinction-and-reinstatement paradigm. Here, rats are abruptly cut off from the drug. They are presented with all the normally associated cues, including the lever and the sounds of pump infusion, but no drug is delivered. Then, several days of withdrawal later, the lever provides the drug again, much to the animal's delight. Reinstatement often causes high lever press counts as the rat slips back into its drug-taking habits.

2. A person's drug use has led to social impairment. Their use may cause interpersonal problems, they may be unable to fulfill their familial obligations, or they may suddenly stop pursuing their hobbies. Obviously, this criterion is quite difficult to model in rats, considering how limited their social environments are in the laboratory setting.* The clos-

* Sorry, Mrs. Frisby. Real rats of NIMH don't experience much pressure to help you.

est approximation would be to compare if rats prefer the drug over a different reinforcer, one famous example being a series of experiments conducted in the 1970s that came to be called "Rat Park." Here, researchers looked at the influence of environment on drug-taking behavior. Rats were assigned to two different housing conditions. One set of animals were housed singly, the standard for lab rats at the time. The other dozen or so animals were put into the Rat Park, a communal environment full of playthings, balls, running wheels, and spaces for hiding. Then all animals were given the option to drink from two different bottles, one containing normal water and the other water with a splash of morphine. The isolated animals drank much more opioid water than the communally housed ones. These findings suggested that social isolation and escalating drug use go hand in hand.

3. A person continues to use the drug even despite the adverse health risks. Debi Austin of 1990s anti-smoking television campaign fame is an excellent example of what this aspect of substance use disorder looks like. In one of her commercials, Austin speaks about her cigarette addiction and how the tobacco industry targeted her and millions of other teenagers. After pleading with the audience not to be the next victim, she takes a long drag from a cigarette through an open hole in her neck, a consequence of a surgery to remove her larynx. In rats, a Skinner box can be designed with electrified floors that deliver a mild but painful foot shock alongside drug delivery. Some rats see physical pain as a powerful deterrent and quickly become abstinent. Others persist despite the knowledge that bad things happen when they go searching for their chemical vice.

4. A person experiences pharmacological changes in response to their drug use. These changes happen at the level of

receptors or cells, and they present as tolerance to the drug or withdrawal from the drug. Caffeine, one of our most innocuous drugs, produces these pharmacological changes: a market exists for both the "trenta," a 30-ounce cup of coffee, and mugs that say "Don't talk to me until I've had my coffee." A chronic user* will require a higher dosage of the drug to experience the effects of that drug. Few people can go four or five rounds of puff-puff-pass between Willie Nelson and Snoop Dogg without falling into complete catatonia. Rats with a history of drug exposure are eager to lever press for additional drug, hinting that they seek the reward more often because the drug effect fades more rapidly, characteristic of tolerance. The other pharmacological change is the withdrawal syndrome, the unpleasant experiences in the absence of the drug that encourages further drug use. Rats also exhibit drug withdrawal. For example, an opioid-dependent rat without the drug shows observable behavioral changes: their unsteady gait causes them to lean up against the walls of the cage for support, they shiver like a wet dog, and they chatter their teeth. They become hypersensitive to pain, sometimes even showing depressive or anxious-like behaviors. In the self-administration setup, they are quick to lever press in the state of withdrawal, just as people often seek the drug to relieve their withdrawal.

These drug Skinner boxes and their modifications have revealed some enlightening tidbits about human drug use, highlighting parallels between rats and people. Spoiler, the answers to the following questions are all yes:

- Are there genetic factors that predispose some animals to take more drugs? Laboratory rats have all sorts of genetic

* Two things that are hard to turn down: good weed and a good double entendre.

differences. Random mutations of the genes that lead to more/less neurotransmitter production or stickier/looser receptors are just two among a multitude of factors that influence brain chemistry after drug exposure. People likewise have random genetic differences. Did you know that natural redheads are more sensitive to opioid analgesics?

- Relatedly, do certain personality traits predict increased drug taking? Some rats are inherent novelty seekers that take gambles with their environment even at the potential risk of dangers. In self-administration terms, rats are given the option to press on either of two levers: One delivers the standard food pellet, while the other delivers three times the amount of food with the added risk of receiving a foot shock at some probability. Risk-takers also take more cocaine in the future.
- Are there times of our lives where the likelihood of drug taking is increased? Adolescent rats behave a lot like teenagers, college students, and twenty-somethings, the age-groups most likely to engage in risky, thrill-seeking behaviors, including initial experimentation with drugs.
- Are there therapies that might help people who are at risk for addiction or death by overdose? Making tiny chemical modifications, such as tacking a proton-hungry fluorine onto an opioid molecule, changes its properties, making it more selective to activation at sites of tissue injury rather than centrally, thus decreasing its misuse potential. Another strategy is to inject an agent that causes the body to produce antibodies that stick onto molecules of heroin in the blood, making the chemical complex too large to cross the blood-brain barrier. Some promising evidence suggests that monkeys with this immune treatment self-administer less opioid.

Two distinct attitudes toward the use of mind-related drugs have manifested themselves in science fiction. One is cautionary. . . . The other is visionary and utopian.

In 1974, the National Institute on Drug Abuse published a book summarizing the body of empirical research on how drug use affects various social and health issues. The first eight chapters cover a range of human experiences all the way from sex and pregnancy through employment and death. Given how much society has changed since disco, most of the conclusions are a historical snapshot of the zeitgeist.

The outlier of the book is chapter 9. Here, the focus isn't on drug-related health outcomes but on overall societal attitudes toward drugs. While the other chapters summarized peer-reviewed studies and clinical observations, there was not yet a major description of this question in the academic literature. To accomplish this, the government contracted science fiction author Robert Silverberg, winner of the Hugo Award and the Nebula Award, to compile a comprehensive list of fictional drugs since the beginning of the century. His bibliography begins in the "primitive period" of science fiction, starting off with the potent utopiate* soma from Aldous Huxley's *Brave New World* (1932). Then, as authors let their firsthand cannabis and LSD experiences fuel their writing, drugs started appearing more frequently throughout the contemporary period of science fiction (1960 to 1970s). Silverberg describes the role each drug played in the fiction and categorized them into nine groups, including the good stuff (euphoriants and consciousness expanders) and the bad stuff (totalitarian mind control drugs and wartime neurotoxins).

* An excellent, perfectly apt portmanteau I had never seen until reading Silverberg's list, combining "utopia" and "opiate."

Around this time when Silverberg and other authors were thinking up interesting reasons why a person might use a drug, scientists were deeply curious about the other question: "What are the neurochemical signals that lead to reward?" An influential scientist named Roy Wise in the late 1960s had a guess it was norepinephrine based on some strong experimental evidence. In a Skinner box, rats learn to lever press for stimulation of their medial forebrain bundle. If these rats are exposed to toxins that block the enzyme that makes norepinephrine, they stop lever pressing. But when you later reintroduce norepinephrine artificially, they resume their lever pressing.

However, some years back in 1957, a tiny four-paragraph article in a top scientific journal, wedged between a discussion of quartz crystal formation and atmospheric density readings collected by *Sputnik 1*, described a brain chemical called 3-hydroxytyramine. Prior to this publication, this substance was already known to be an intermediary compound, a biochemical stepping stone along a series of other chemicals ultimately leading to epinephrine, the chemical "go" button that fires up the sympathetic nervous system and prepares an animal for action. But this publication hinted that 3-hydroxytyramine had some interesting features on its own. This was one of the first entries that drew attention to a chemical that has captivated neuroscience research and flooded public discourse for decades. Chemical names are cumbersome and not very memorable, so "3-hydroxytyramine" held interest only for hard-core biochemists. Most of us, however, know this as pop culture's trendiest neurotransmitter: dopamine.*

The data started to shift away from norepinephrine as being the key to reward. Wise showed that if you block the dopamine receptors using the chemical pimozide, rats eat significantly less food

* The lead author of that short article, Arvid Carlsson, went on to win a Nobel Prize in 2000 for his pioneering work on dopamine.

but only after getting an initial taste of the food. This suggests that the food no longer provides the reward it is supposed to provide. Wise concluded this was analogous to the human symptom anhedonia: a lack of reward. Based on the involvement of dopamine in these studies, dopamine came to be thought of as the pleasure neurotransmitter. According to this hypothesis, substances that increase brain levels of dopamine have the potential to be habit forming since they produce pleasure. This idea became so influential that pharmaceutical companies would routinely measure the dopamine-boosting effects of their new compounds and discard those that did for fear that these drugs may be addictive.

Then a disruptive force by the name of Wolfram Schultz challenged the dopamine hypothesis. Schultz and his team wanted to study what the dopamine neurons were doing in real time as an animal was exposed to rewards under different conditions. His team of scientists buried tiny electrosensitive probes deep into the ventral tegmental area (VTA) of monkeys, which could single out the firing properties of individual neurons, hence providing a cellular correlate for the dopamine signal. Initially, some of the neuronal data agreed with Wise's hypothesis. If you position a straw in front of the monkey and randomly deliver a tiny droplet of delicious apple juice, the dopamine cells send off a burst of action potentials in response. Here, more dopamine = more happiness.

The next sets of experiments changed our understanding of dopamine. Instead of the occasional, unexpected treat, what happens if the animal is presented with a cue that signals the delivery of juice? With a few pairings of light followed by water, the animal learns to associate the two, just like Pavlov's dogs salivating at the sound of the bell. The dopamine cells start firing when the lights turn on. But once the liquid hits, those cells no longer increase their activity like they originally did. Schultz got dopamine cells to fire not on receiving something rewarding but simply by presenting cues associated with reward, contrary to the predictions offered by the previous hypothesis. Dopamine tells us to get ready because something good is about to happen.

Now flash the light on but deny the animal access to the reward. The dopamine signal rises in response to the light, but it decreases at the moment when they would normally expect to get the reward. Here, dopamine tells a complicated story. The spike in response to light says, "Here comes the good stuff!" But the silence says, "Hold on, something's wrong. Where's my treat? Did you do something different? Figure it out so you can get a hit of that delicious water." Based on this convincing new evidence, Schultz proposed a more complex theory about what dopamine does for us. According to the reward-prediction error hypothesis, changes in dopamine levels are responsible for making us aware of when our expectation of the world doesn't align with the outcome. We keep doing the things that lead to an increase in dopamine spikes. Inversely, we avoid the things that lead to a silencing of that signal. Whenever there is a mismatch, we update our understanding of our surroundings. Dopamine doesn't just tell us what is good or when to be happy. It's evolution's best teacher.

So consider the reward-prediction error hypothesis in terms that our forager ancestors could easily relate to. Eating a sweet berry for the first time would be a good experience, as it provides the body with precious calories (an unexpected reward produces a burst of dopamine). Whenever they would catch a whiff of that delicious fruit tree again while searching the plains, they would know there is a source of more fruit nearby (the scent, which is a learned stimulus that predicts reward, produces a dopamine spike). However, reaching the tree only to discover that it has already been picked clean would be a tremendous disappointment (stimulus but no reward leads to a dopamine dip).

Most recreational drugs interfere with dopamine signaling albeit through different cellular mechanisms. Nicotine activates excitatory receptors found directly on the dopamine neurons, causing them to fire more often. Cocaine and amphetamines inhibit the dopamine reuptake mechanism, which prolongs the presence of dopamine in the synapse, resulting in increased signaling. Alcohol, barbiturates, and opioids activate inhibitory cellular mechanisms, prompting the obvious question of how cellular inhibition can produce a contradictory excitation of dopamine cells. In the VTA, there are other types of neurons, such as inhibitory cells that are steadily clamping down on the amount of dopamine being released. When the receptors on these cells become modified by ethanol or morphine, the cells fire less, causing the dopamine neurons to fire more by disinhibition. Direct excitation by nicotine is stepping on the gas, while disinhibition is easing off the brakes as you roll downhill.

Whenever a drug drives up the dopamine signal, the brain reacts as if it were a primary reinforcer, such as food or water. It wants more. In theory, the person gets the dopamine signal whenever they are simply exposed to the cues associated with drug use, just like the light in Schultz's rat cages. These cues include physical sensations, such as the feeling of tapping the cigarette pack against the palm or the sulfurous skunky smell of good ganja. They also include abstract cues, such as a social vibe from the people who are always around on binge-drinking nights or the set of emotions accompanying a drive to a trap house in a different part of the city.

> In his fantasy number he was driving past the Thrifty Drugstore and they had a huge window display; bottles of slow death, cans of slow death, jars and bathtubs and vats and bowls of slow death, millions of caps and tabs and hits of slow death, slow death mixed with speed and junk and barbiturates and psychedelics, everything—and a giant sign: YOUR CREDIT IS GOOD HERE. Not to mention: LOW LOW PRICES, LOWEST IN TOWN.
> —Philip K. Dick, *A Scanner Darkly*

Much of *A Scanner Darkly* revolves around a fictional drug called Substance D, the "slow death" of this character's imagination and the object of his fixation. Substance D is derived from a plant, as many recreational drugs are in reality. For example, nicotine is a potent insecticide, found in tobacco and the leaves of tomato and eggplant. Poppy, the source of opium, was cultivated as far back as the ancients, with references to its amnestic properties dating back to Homer. The energizing leaves of the coca plant have been chewed on and tucked into the cheeks of precolonial South Americans for centuries before laboratory scientists isolated cocaine. But not all psychoactive compounds come from plants.

Consider the story of ergot, the common name of *Claviceps purpurea*. This fungus grows in little black kernels alongside rye, wheat, and other agricultural grains. If a farmer harvests the grain without separating the fungus and the miller grinds up this impure flour, the villagers end up with tainted bread. For the lucky ones, nothing happens. The others develop ergotism, a type of poisoning characterized by convulsive symptoms, gangrenous symptoms, or both. Convulsions, the uncontrollable muscle contractions leading to twitching and muscle cramps, are unpleasant at best and painful at worst. Gangrene, the loss of blood flow so severe that body parts turn black, die, and literally fall off, is obviously bad. Surely, there are no medical applications for something convulsive and gangrenous, are there? As science fiction pioneer William Gibson will eventually write, "The street finds its own use for things."

Giving birth was one of the more dangerous activities for women of the Middle Ages. One of the biggest risks for both peasants and nobility alike was death by rapid blood loss. Childbirth happens to be one of those rare moments when you want convulsive and gangrenous symptoms. If you give ergot, intensity and the frequency of uterine contractions increases while also decreasing blood flow in the aftermath, and you have shorter, less strenuous labor coupled with a safer outcome.

Many years later, English settlers new to the American colonies discovered wide swaths of wild rye, which grew particularly well in the Atlantic coast climate. By early 1692, people in villages and settlements were getting sick. They didn't consider a rogue fungus baked into their tainted rye bread as the cause of their illness. They didn't correlate that the previous year brought early rains and a hot summer, ideal growing conditions for fungi. Instead, they blamed something more in line with their beliefs and understanding of the world.

Witches. Satan was loose in Salem.

When the trials began, the victims were asked to describe how they knew they were "bewitcht." Some described painful muscle spasms, as if their bowels were being pulled out of their bodies, while others fell into epilepsy-like seizures, both symptoms consistent with convulsive ergotism. Yet others described sudden sensations of being pricked with pins or being bitten, akin to the feeling when your foot falls asleep, which is in line with gangrenous ergotism. An illness from the food was being leveraged as a weapon to prosecute and punish freethinking, ethnically different, or otherwise unlikable women.

Much more interesting than the physiological changes are the brain changes that accompany ergot poisoning. Some people experience confusion, while some enter a manic or psychotic state. Others have vivid hallucinations. Art historians suggest that Dutch Renaissance painter Hieronymus Bosch may have been inspired to paint his most famous triptych based on ergot-driven hallucinations. The critters that populate his *The Temptation of Saint Anthony** are difficult to describe in words, but until you see a picture of it for yourself, "voyeur wolf-faced fellow" and "ashy nun

* Ergotism was common enough that an entire order of monks, the followers of St. Anthony the Great, dedicated part of their work to treating these patients. Ergotism came to be called "St. Anthony's Fire" in reference to the burning sensation that results from the loss of blood flow.

swordbaby in a basket" will have to suffice. There are fish boats and jug horses and jester dogs roaming the countryside while the holy saint himself kneels in the center of the oil painting, nonchalantly going about his prayers as chaos envelops him.

But the story of ergot is only at its beginning. In 1938, a Swiss chemist named Albert Hofmann was tasked with finding a process to refine a heart medication. In the laboratory, he applied the process of hydro*lysis* to isolated *erg*ot compounds, making the product lysergic acid. Although this drug held moderate promise, it had limited practical application because it degraded very easily. He developed more than 25 different modifications of that compound, hoping to find a more stable version. However, none of these compounds was the breakthrough drug he was hoping for, so many of them sat on a shelf, largely forgotten.

Until five years later. Hofmann decided to revisit these synthesis experiments on a mid-April Friday when he unexpectedly had to leave the laboratory. He was feeling ill. He lay down on his couch and closed his eyes as he had done many times before. On this day, however, he found the bright sunlight to be overpowering. Drawing the shades for a little peace, he closed his eyes and was soon greeted with an "uninterrupted stream of fantastic pictures, extraordinary shapes with intense, kaleidoscopic play of colors." With the weekend to reflect on what had happened, he figured that he must have accidentally touched some of the compound that caused this illness. Of course, on returning to work, he did what any other scientist would do.

Drink it.

On Monday, April 19, 1943, Hofmann prepared himself a beverage. He measured out 0.25 milligrams* of his lysergic

* As someone who has regularly measured tiny crystals in a laboratory, accurately measuring anything smaller than one milligram is extremely difficult.

acid diethylamide-25 salt crystals and dissolved the tiny flakes in water. With a quick gulp, he downed the vial. This was all unexplored territory, so he had no idea how much to take. If the strange sensory experiences didn't repeat, maybe he'd try again with a higher concentration.

On this day, he gave himself 10 times a normal hit of LSD: acid.*

Like the well-trained scientist he was, he took careful notes on the whole experience, recording the exact time that he began the experiment. He rode home by bicycle, as wartime restrictions had put limitations on automobile use. Now in the comfort of his home, he sat waiting to see if this drug was the cause of his twisted visual distortions from last Friday. Within hours, he was hit by a "most severe crisis." In his own words:

> *My surroundings had now transformed themselves in more terrifying ways. Everything in the room spun around, and the familiar objects and pieces of furniture assumed grotesque, threatening forms. They were in continuous motion, animated, as if driven by an inner restlessness. The lady next door . . . was no longer Mrs. R., but rather a malevolent, insidious witch with a colored mask.*

An overwhelming out-of-body experience consumed his existence: His body was there, but his mind was completely detached from reality. He felt as if he would die, embarrassingly at the hands of his own creation. Luckily, the panic of this drug-induced self-experiment dissipated overnight. The next morning, he found a renewed view of the world. His breakfast tasted better than ever. He strolled through the garden with an appreciation for nature's beauty, dewdrops glistening in the sunlight. The world was reborn.

---

* Mistakes in the lab, a track record of self-experimentation, and the occasional megadose of a very potent psychedelic aside, Hofmann lived to be 102 years old.

Looking back on his trip, Hofmann was quick to recognize the value of a drug that can reshape the mind and all its experiences. Later experiments found that a mere 0.02 milligrams was enough to produce a psychic effect and a slight euphoric state minus the ominous thoughts of death that haunted Hofmann. It was marketed as Delysid and sold to psychoanalysts who prescribed hits of acid to help encourage breakthroughs with their patients, particularly to bring up repressed memories for discussion. It was used to help people give up alcohol. Grant applications, scientific publications, and academic conferences popped up to more fully explore the benefits of LSD and similar mind-expanding drugs in psychiatry.

But society at large had different ideas for the drug. Some people enjoyed the euphoria and the weird perceptual changes of an acid trip. It became widely used as an inebriant as people began to, in the words of Timothy Leary, Harvard professor and one of the most influential psychonauts of the counterculture movement, "turn on, tune in, drop out." Hofmann, seeing his most promising compound being misused by the citizenry, started referring to LSD as his "problem child." The CIA bought up most of the world's supply of acid and started conducting human experiments using the code name Project MKUltra under ethically abhorrent circumstances. In a knee-jerk backlash against the counterculture movement, the drug was made a Schedule I drug in 1970, the highest and most stringent level of regulation. After this, funding for LSD research dried up, as did many of the labs that were working to figure out how the drug works and how it may be helpful for therapy.

Some 30 years later, research into psychedelics started to recover. The current renaissance has its roots at the Johns Hopkins Center for Psychedelic and Consciousness Research, but several hospitals and research institutions around the world are actively recruiting participants for studies, many of them government funded. Small trials have demonstrated applications for treating a wide variety of psychiatric illnesses, including depression, anxiety, obsessive-compulsive disorder (OCD), eating disorders, and even tobacco and alcohol use disorder. Any application in psychotherapy is justification enough to reschedule the drug, easing legal restrictions on production quotas that have increased the cost of doing research. In fact, many cities, including San Francisco, Ann Arbor, Seattle, and Minneapolis, have passed decriminalization bills for psychedelics, following in the footsteps of Denver, the first city to do so in 2019.

Much work has yet to be done on how LSD, psilocybin from magic mushrooms, N,N-dimethyltryptamine (DMT) from ayahuasca brew, mescaline from the peyote cactus, bufotenine from the Sonoran Desert toad, and other psychedelics exert whatever benefit they might offer. Most of the attention is on the serotonin receptors, as all these chemicals activate those receptors robustly, particularly the 2A subtype. But there is still quite a chasm between the molecular excitation of neurons and the total ego dissolution that leads to a complete spiritual overhaul when under the influence of one of these substances. Ever since the 1960s, psychedelics were likened to being "a chemical key" that unlocks the mind, an idea supported by modern functional imaging studies. Our brains have a series of structures collectively called the default mode network (DMN) that are proposed to be essential for identification of the self. This network is always active while the person is awake but not focused on any mind-intensive task. When the DMN is shut down in the presence of psychedelics, so does the person's self, producing that feeling of ego dissolution. The DMN is akin to deep ruts in the grass as a wagon is pulled between settlement towns. Psychiatric disorders can be likened to repeat back-and-forth travel between those towns, each persistent thought digging the ruts a little deeper: I am unworthy of love, I must have my morning cigarette, I must put these items in their exact place, and so on. Psychedelics add a little bit of chaos to the brain system and kick the wagon out of the rut.

But the man who comes back through the Door in the Wall will never be quite the same as the man who went out. He will be wiser but less cocksure, happier but less self-satisfied, humbler in acknowledging his ignorance yet better equipped to understand the relationship of words to things, of systematic reasoning to the unfathomable Mystery which it tries, forever vainly, to comprehend. —Aldous Huxley, *Doors of Perception*

CHAPTER FIVE

THE BENE GESSERIT TEST

The Story of Pain

> Pain is simply our intrinsic medical adviser to warn us and stimulate us. —H. G. Wells, *The Island of Doctor Moreau* (1896)

LIKE ALL THINGS NERVOUS SYSTEM RELATED, THE PAST CENTURY has been one of rapidly expanding frontiers. In 1900, Charles Scott Sherrington first defined nociceptive pain, from the Latin *nocere*, meaning "to harm." This is the protective pain that comes from damaging, injurious stimuli. In 2021, the Nobel Prize was shared by David Julius and Ardem Patapoutian for their work describing the molecular mechanisms of scalding heat, freezing cold, crushing pressure, and the face-melting experience of biting into a Carolina Reaper, one of the spiciest peppers in the world.

The physical perceptual component of pain is best understood through the lens of molecular and cellular neuroanatomy. It starts in the periphery with cells called nociceptors, first-order neurons that are specialized for sensing pain. These serve as the bridge between the skin when it functions as a sense organ and the central nervous system. They are among the longest neurons in the body: a first-order neuron, feeling the pain from stubbing your toe on a wooden bed frame that hasn't moved in years, extends all the way to the base of your spinal cord. These neurons have their cell bodies in ganglia, one of several clumps of cells found outside of the central nervous system. They send two sets of axonal projections. One of them reaches toward the periphery, forming branches close to the skin's surface called free nerve endings. Here, the neuron expresses a variety of receptors, some of which include the following:

- TRPV (pronounced "trip-V") receptors are responsible for sensing if we were the victim of Charlie McGee's pyrokinesis. The feeling of noxious heat, temperatures in excess of 109°F (43°C), excite these nociceptors. Interestingly, it is the same reason that ghost pepper salsa produces a burning sensation in our mouths. Capsaicin, the key molecule found at distressingly high levels in hot peppers, is an agonist that binds to the TRPV1 subtype of the receptor. Alcohol also modifies these receptors by lowering their threshold for activation to well below body temperature, contributing to the warming effect of a shot of strong booze.
- TRPM receptors sense the end of the thermal scale, the freezing chill felt by the crew of the Antarctic expedition team in *The Thing*. One specific subtype, the M8 receptor, also responds to menthol, the reason why peppermint and Tiger Balm produce a cooling sensation.
- Piezo1 and Piezo2 receptors are mechanically sensitive channels that detect the feeling of being under the curse described in Junji Ito's *Splatter Film*. The experiments that demonstrated Piezo receptor activity are technically very challenging. Single cells were gently pushed from the outside, one micron at a time, while changes in their electrical properties were detected. The stronger the poke, the greater the change.

- TRPA1 receptors are the ones that drive up nociceptor activity when you get caught in a cloud of noxious tear gas or, more likely, when you eat wasabi or horseradish. These receptors are found along the same population of neurons as the TRPV receptors, so a lot of these gaseous agents/condiments produce a burning sensation by activating the same neural pathway.

The variety of receptor types account for why there are many different flavors of pain, each of them best described not by the singular word "pain" but by adjectives. There is the gradual stinging burn from slicing a jalapeno before rubbing your eyes, the sharp transient prick of a zap of static, the deep throbbing heat of muscles the day after a long workout, and the rhythmic pulsing of a pesky headache. My favorite descriptions of pain are those written by the late Justin Schmidt, an entomologist who dedicated his life to studying and experiencing the many stings and bites of wasps, bees, and ants. He writes like a sommelier of pain, savoring many of Mother Evolution's finest blends of venom-induced agony: "A flaming match head lands on your arm and is quenched first with lye and then sulfuric acid" (the western honeybee), "Tiny blowtorches kiss your arms and legs" (the yellow fire wasp), and "Like walking over flaming charcoal with a three-inch nail embedded in your heel" (the bullet ant).

The other branch of the nociceptor neuron forms synapses on other neurons within the spinal cord. One synapse may form onto motor neurons, which project back outward toward the muscles, allowing us to perform reflexive actions, such as a rapid withdrawal from something sharp, even without conscious involvement of the brain. They can also synapse with the next neurons in the chain of signaling upward, called second-order neurons. These cells send their axons toward the brain, eventually synapsing in the ventral posterior nucleus of the thalamus. Here are the third-order neurons, which in turn signal further upward for processing through higher-order cortical circuits. Once a signal makes its way into the brain and gets processed by those circuits, it transforms from simple nociception to complex pain. This last step is the real black box.

Dune pun intended.

> A human can override any nerve in the body. —Frank Herbert, *Dune**

The nociceptive pathway is only part of what makes up the experience of pain. Pain is so much more than just some ancient circuit of neurons to protect the body from harm. There's also a complicated cognitive and emotional element to the experience. Pain is a high-order process modified by sociocultural elements. And to demonstrate this point, consider the following examples:

- A woman in a monogamous marriage who loves and cares for her husband in every way restrains him and whips him on the back and genitals in their self-described sex dungeon. It has become a necessary element of their sexual encounters as part of foreplay. Sure, they could see a psychiatrist if either of them had distress about their desires. But a diagnosis of sexual sadism disorder is unlikely to change any aspect of their happy and fulfilling sex lives.
- A 34-year-old man breaks his hand, then continues intense physical activity without slowing down. As the underdog in his first heavyweight fight, Tony Bellew walked away

* The Bene Gesserit are an order of women with mystical powers and political influence. At the beginning of *Dune*, the protagonist is tested by the Reverend Mother, a Bene Gesserit who threatens him with the following predicament: place your hand into the box, and you will feel tremendous pain. If you pull your hand from the box, you will die.

victorious. In a post-match interview, Bellew said, "I don't feel the pain, all I think about is winning." During this same fight, his opponent, David Haye, tore his Achilles tendon in the sixth round, an injury so severe that you can sometimes hear a sickening "pop" the moment the tendon snaps. Haye fought through the pain until the eleventh round, when his trainer threw in the towel.

- In act III of Tchaikovsky's *Swan Lake*, the dancer who plays the role of Odile, the black swan,* has a moment's rest before performing one of the most technically challenging sequences in classical ballet: 32 *fouettés* in a row. Not only does this require putting the entire body weight on one toe, but it also uses a violent throwing of the other leg as the sole means of propulsion. *Fouetté*, after all, comes from the French word meaning "to whip." Despite the bruises, stress fractures, ingrown toenails, and blisters that are all an expected part of this art form, this is one of the most coveted roles in all of ballet.
- In a reenactment of the Passion of Jesus Christ, a man carries a heavy wooden cross through the street and is later crucified as six-inch steel spikes are nailed through his hands and feet. He remains in this torture for up to 30 minutes, taking ragged shallow breaths beneath the humid, tropical air. Annually, up to 20 people across the Philippines undergo this brutal form of enlightenment; Ruben Enaje has done it 35 times. After the ordeal, he gets checked by medical staff, rests briefly, and then walks home on now holier (and also hole-ier) feet.
- Rites of passage are a cross-cultural tradition, but that of the Sateré-Mawé of the Amazon is quite different. In

* The duality between the white and black swans and the challenges of handling this role physically and mentally are the central focus of the psychological horror film *Black Swan*.

> preparation for the ritual, village elders wrangle up swarms of venomous ants and drop them into an anesthetic brew. They are then weaved, stingers first, into palm frond mittens. For a man to become eligible for marriage and leadership in the community, he must don these gloves for 30 minutes while these terrified ants deliver successive doses of toxin all over their hands. By the way, these are bullet ants (see Justin Schmidt's description above for an idea of why their sting is given a 4+ on his pain rating scale that goes from 1 to 4). Once the boy recovers from days of hand paralysis, nausea, cardiac arrhythmia, and uncontrollable shaking, he can be satisfied he is 5 percent closer to being a man (he does it 19 more times).

Pain is so much more than simple nociception.

After the third-order neurons get modified by thalamic processing circuits, their outputs project into the cortex, which interprets the pain. Human imaging studies have identified which of these higher structures are then involved. Unsurprisingly, the somatosensory cortex, the strip of the parietal lobe that handles bodily perception, is a major player. These circuits localize stimuli, telling you approximately where on your skin the pain is coming from. There are far more somatosensory neurons dedicated to processing pain in your hand compared to the skin on your back, for example, which is one reason why you can precisely identify where a splinter in your finger is down to the millimeter.

Another cortical circuit recruited during pain is the anterior cingulate cortex. These structures are less concerned about where the pain is. Rather, they form the emotional component that comes with pain, creating a link between the physical experience and the more abstract affective feeling. When put into a scanner while the forearm is exposed to increasingly hotter temperatures, the anterior cingulate increases its activity, as if to say "*Aargh*, these scientists are the *worst*!" This interlacing between the physical and emotional components of pain manifests in idiomatic language. People can hurt your feelings, unrequited love can cause heartbreak, and any movie where the dog dies is gut-wrenching. Activity through the anterior cingulate is likely the reason behind this overlap. To model socioemotional pain, participants were asked to play a digital game of catch with two other players. An avatar that represents the participant is displayed on a screen. The avatar would receive the ball, after which the participant can select who they want to pass the ball to. At first, the ball would get passed between all participants with equal probability. At some point, however, the two other players, who are not actually real people, start to exclude the human, creating an artificial and wholly abstract sense of exclusion. This manipulation fires up the anterior cingulate even though there is no physical stimulus throughout this entire experiment, suggesting that the emotional pain of social exclusion is a lot like physical pain, at least to the brain.

So far, the circuitry described doesn't have many useful applications for a person trying to better manage their pain when being tested as the prophesied Kwisatz Haderach, while participating in a brutal rite of passage into adulthood, or after you run over a toe with a vacuum. But there are a few tricks out there based on our neuroanatomical wiring plan and the workings of human psychology.

The most basic, physical sensation of pain can be dampened by preventing signals from going upward. Both nociceptive and non-nociceptive first-order neurons converge at the spinal cord, where both populations synapse onto second-order neurons. Non-nociceptive neurons carry information about other somatosensory modalities, such as crude touch and vibration. If you overwhelm the second-order neuron with non-nociceptive inputs, the second-order neuron cannot determine if the incoming information comes from something painful or something innocuous.

This idea, gate control theory, accounts for why rubbing a body part after injury helps soothe the pain.*

The ascending neural pathway is just one part of the story. There's an equally important reciprocal pathway that projects downward, called the descending pain modulation system. This circuit has its origin at the periaqueductal gray (PAG) of the midbrain. The output of the PAG is further downward in the raphe nucleus of the medulla. These neurons send their long axons down through the spinal cord, where they release serotonin at synapses onto interneurons, resulting in decreased communication between the first- and second-order neurons. Activating the PAG leads to a less intense upward nociceptive signal, thus decreasing pain.

Many people don't have much conscious control over the circuits of the PAG. One man, however, might. Dutch athlete Wim Hof is best known for a handful of superhuman accomplishments. He ran a marathon at the Arctic Circle wearing only shorts. He climbed Mount Everest without a shirt. He sat in an ice water bath for 72 minutes. These and other cold-related stunts have earned him several Guinness world records and his nickname: the "Iceman." In an imaging study, the Iceman and control participants were put into a suit lined with plastic tubes through which water of different temperatures could be passed, alternating between warm (93°F, or 34°C, a pleasant warmth) and cold (59°F, or 15°C, a wholly unpleasant chill on the skin). In the cooling condition, the activity in the Iceman's PAG increased. This suggests that his ability to withstand such painful conditions is dampened by some voluntary control he might be exerting over his PAG.†

---

* In a tangentially related life hack, pain and itch are carried by the same set of fibers. Gate control theory suggests that slapping an insect bite abates itch just as well as scratching, with the bonus that slapping produces less tissue damage and inflammation.

† Legal issues have landed Hof, ironically, in hot water. Claims that his techniques can alleviate symptoms of autoimmune disorders are unsupported by experiments, casting a shadow of pseudoscience over his methods. Furthermore, a $67 million lawsuit claims that hyperventilation and cold-water submersion, two aspects of his technique, led to the drowning death of one such practitioner.

Although many of us do not have conscious control over the PAG, we do have some capacity to inhibit pain through activation of higher-order cortical systems. Attention plays a major role in pain. When pain is working in its protective capacity, the way it's meant to, your attention is immediately drawn in by pain. The inverse is also true: the more your attention is drawn away from the painful stimuli, perhaps by some difficult mental tasks that engage the prefrontal cortex, the less pain you experience. For example, while performing an n-back task, a cognitively demanding memory test that requires the maintenance of a string of digits while occasionally reporting back the number that appeared several digits back, people report less pain. In another task called the Stroop task, names of colors are shown to the participant. But they appear in a color that is different from the text, so the word "blue" might appear in red. The participant naturally wants to read the word, but the correct response would be to say the color of the word. This task requires intense suppression of a response. As with the n-back test, people doing the Stroop task report less severe pain.

But my favorite painkiller is the proper use of improper language. In possibly the only appearance of the word "fuck" in the academic literature,* participants were able to keep their hand submerged in a bucket of ice water for longer if they chanted the f-bombs like a meditation mantra. As a control condition, they were asked to repeat either their preferred adjective used to describe a table or the humorous and vaguely insulting "twizpipe." Mechanistically, it was proposed that the use of curse words activates an emotional response that helps decrease the sensation of pain through autonomic nervous system activity. In line with this hypothesis, the analgesic effect is related to how often a person

* There is a nonzero chance that my work account has been flagged by the information technology team. Granted, this is one of my more innocuous searches, especially in comparison to other queries, such as "How deadly is rabies virus?" and "How expensive is cocaine?"

uses profanity in everyday life. It has a stronger effect if people are generally pure of speech since these words produce a stronger emotional response. But for those of us whose daily speech is a regular string of profanity, the pain dampening effect is less potent.*

> Writers remember everything . . . especially the hurts. Strip a writer to the buff, point to the scars, and he'll tell you the story of each small one. From the big ones you get novels. A little talent is a nice thing to have if you want to be a writer, but the only real requirement is the ability to remember the story of every scar. —Stephen King, *Misery*

Pain is a remarkably effective warning mechanism. Broken bones, meningitis, appendicitis, and many other illnesses can have fatal consequences if not caught early, and pain is usually a sign that something is amiss. Unsurprisingly, pain is the principal complaint that would lead someone to the doctor. In addition to this bodily "check engine" light, pain is an amazing teaching tool. Local inflammation leads to hypersensitivity through the action of signaling molecules released during injury, reminding us to be extra careful and thus protecting the site from further injury or infection. In addition, it helps us avoid future pain by instilling within us long-lasting lessons collected over the course of our life about which cues predict pain. Enough pine needle pokes, enough little wooden splinters, enough accidental sticks of a steak knife, and you learn which parts of a cactus or a porcupine to avoid based on shape alone even though you may have never encountered these things before.

* This groundbreaking research team was awarded with the Ig Nobel Prize, an award given for somewhat silly findings. Other winners include entomologist Justin Schmidt for serving as a human pincushion to bugs.

People commonly attribute learning to the hippocampus. Since pain elicits a strong emotional response, there is no surprise that neural pathways through both hippocampus and amygdala are involved in pain memory formation. However, there is a subtle nuance to which types of pain-associated memories are involved in each of these two unique but closely related circuits. Consider a study in which rats learned different types of pain associations. First, they learned to pair a specific environmental context, such as a mysterious new cage with a unique odor and parallel metal bars for floors, with an inescapable foot shock. Second, they learned to associate an unnatural cue, an 800-Hz sine wave tone,* that predicted the inevitable foot shock. After just one pairing, rats learned to anticipate the pain, freezing in their tracks. When the amygdala is lesioned, the rats fail to form any associations, continuing to walk around the cage even when a tone is played, blissfully ignorant of the upcoming electrical current about to be passed through its paws,† But when the hippocampus is zapped away, the rats learned only the link between the sound cue and the pain, not the place context and pain. In human terms, both the hippocampus and the amygdala tell you that the smell of the dental office may suggest pain is near, but it is only the hippocampus that reminds you that the whirr of a drill precedes pain.

Since it is well documented that memory is fallible, there is no reason why pain memory isn't also subject to a little editing. One such trick was demonstrated during a procedure that is a source of apprehension for many men approaching *that* age, 45, when the American Cancer Society recommends the first

* 800 Hz is a weird note that is not in the conventional Western musical scale, somewhere in between a G and G#.

† This procedure is called the fear conditioning paradigm and is most often used as a nonhuman model for studying posttraumatic stress disorder (PTSD) rather than pain, reminding us that "all models are wrong; some are useful."

colonoscopy. This anxiety comes from the pain and discomfort of having a tube several feet long inserted through the anus. To ease the process, patients are often given a mild opioid and a sedative, but still, people typically report a pain score somewhere between 4 and 7 during the process. In one particular study,* patients were randomly assigned to two groups. One received the standard colonoscopy procedure, while the others had a modified procedure that included an extra step at the end where only the tip of the probe was left in the rectum for an additional minute. This step is significantly less painful than any other part of the procedure, but it causes the procedure to last longer altogether, averaging closer to 28 minutes.

According to self-report scores collected during the procedure, the pain experienced was equal between groups for the beginning and middle of the probing, with decreased pain only at the end for the modified procedure group, as predicted. However, when polled about their pain experience after the influence of the analgesia wore off, people rated the modified procedure as overall less painful even though it was longer. Importantly, those who underwent the modified procedure were more likely to return for the recommended follow-up repeat procedure years later. Since aversive experiences in the clinic decrease the likelihood that a patient will return, the recommendation is for health care providers to introduce gentle procedures at the end of medical examinations that let the pain subside gradually.

Most of the time, pain is helpful. But in the following examples, it's a pain in the ass.

---

* The senior author on this study was Nobel Prize–winning economist Daniel Kahnemann. Academia is full of journals with bloated titles like *Proceedings of the National Academy of Sciences* and *Annals of Clinical and Translational Neurology*. It's quite a breath of fresh air to see the name of the journal in which this study was published, simply called *Pain*.

> The first time I hooked up to a kid it was like I'd never felt pain before. Even tuning in the pinprick was excruciating. It was pure, unadulterated, hi-fi pain. Jesus Christ! Diaper rash alone felt like a second-degree burn, and the double hernia the kid had I could not believe. I saw colors. I nearly passed out.
> —Valda Peach, *The Pain Addict**

Nociceptive pain protects the body from damage. But there are many examples of when pain serves no purpose. This may result from acute injury, repeat strain, diabetes, multiple sclerosis, migraine, and the unlucky possibility of just being born that way. Non-nociceptive pains go by many names in the parlance of diagnosis: carpal tunnel syndrome, fibromyalgia, and complex regional pain syndrome, to name a few. One study suggests that chronic pain affects up to 30 percent of the population worldwide, leading to lost productivity, a severely worsened quality of life, depression, and opioid misuse. If nociceptive pain is that ear-piercing alarm that gets everyone safely out of the burning building, maladaptive non-nociceptive pain is *The Boy Who Cried Wolf*.

One culprit for the condition is the gene SCN9A, which codes for a protein called $Na_v1.7$, a type of voltage-gated sodium channel found on first-order nociceptive neurons. These channels are one of the main contributors for the regulation of excitation, determining when and how often a neuron can fire action potentials. Normally, these channels have a biomechanical mechanism that prevents too much excitation, a physical blockade that limits their spiking rate. Nociceptive neurons should be silent without noxious stimuli to

* Technology becomes available where physicians can feel what their patients feel, allowing them to more accurately diagnose illnesses. This story was later adapted into TV in a *Black Mirror* episode (2017). The inspiration was drawn from a moment from the author's life while they were injured in a Spanish hospital with an insurmountable language barrier separating Peach from treatment. If you're not familiar with Valda Peach's other work, you would probably recognize him better without his pseudonym as Penn Jillette of the magician duo Penn and Teller.

activate them. But when a neuron lacks the speed-limiting mechanism on its sodium channels, it sends a barrage of action potentials when it should send only a few. This accounts for why people with allodynia, the sensation of pain during exposure to non-painful stimuli, are tempted to shave their heads bald: whenever a stray hair brushes gently across their forehead, they feel a fiery, burning scraping sensation, akin to being tattooed.

More than just hyperactivity of pain neurons in the periphery, some types of pain come from the brain. Phantom limb pains, for example, are very real even if the limbs are not. As the revenge-obsessed Captain Ahab says about his whalebone peg leg, it is "one distinct leg to the eye, yet two to the soul." Many describe their phantom limb pains as a crawling feeling on the skin, a burning, the sudden feeling of a spike driven through the missing body part, or the feeling as if their hand is clenched and won't open. Up to 85 percent of amputees have these sensations, adding another challenge to the quality of life beyond adjusting to limb loss. Every body part is sensitive to the phenomenon, such as phantom erection pain following a penectomy. Internal organs are also prone to phantom sensations. After a hysterectomy, people may feel period cramps, and after a rectal amputation, people can feel phantom "pharts."

Warfare has always been ugly, no matter how civilized a society pretends to be. Penetrating injuries, especially those caused by bullets, are particularly nasty wounds. Whenever a bone is struck, it shatters into several tiny pieces, all of which need to be removed. There have always been battlefield surgeons with the skill to perform these delicate operations. But in those centuries without anesthesia and with a line of similarly wounded patients screaming in agony, time spent in surgery was best kept to a minimum. Simply cutting off the wounded body part was in the best interest for both the patient and the army. Unsurprisingly, it was a war surgeon who made the earliest Western descriptions of phantom pains. In the 1500s, a time in medicine when amputation stumps were cauterized by a dunk in boiling oil, Ambroise Paré wrote that "patients who many months after the cutting away of the leg grievously complained that they yet felt exceeding great pain of that leg so cut off."

Sticking with the theme of war surgeons being at the forefront of amputation science, Civil War doctor Silas Weir Mitchell* first put a name to the experience, calling it a "phantom of the missing member, a sensory ghost." The experience was described by Mitchell in a case study in *Atlantic Monthly*, a popular press periodical covering a variety of intellectual topics of the day. He wrote about the recovery of George Dedlow, a soldier serving in an infantry regiment who underwent amputation surgery following a severe injury. On behalf of Dedlow in first person, Mitchell writes,

> *I tried to get at it to rub it with my single arm, but, finding myself too weak, hailed an attendant. "Just rub my left calf," said I, "if you please."*
>
> *"Calf?" said he, "you ain't none, pardner. It's took off."*
>
> *"I know better," said I. "I have pain in both legs."*
>
> *"Wall, I never!" said he. "You ain't got nary leg."*
>
> *As I did not believe him, he threw off the covers, and, to my horror, showed me that I had suffered amputation of both thighs, very high up.*

Mitchell chose *Atlantic Monthly* for two reasons. First, it had a wider audience compared to esteemed medical journals, such as the *New England Journal of Medicine* or *Lancet*. Surgeons who were familiar with Paré already knew of such sensations, but he was more interested in bringing that information to a broader population. Second, Dedlow wasn't real. He was a fictional person

* Presumably different from the actor with the same name, although no one has ever seen them in the same room together.

who represented an amalgamation of the people whom Mitchell had treated, sort of an average of their experiences. However, the public wasn't aware that Dedlow was a hypothetical patient, and the generous readership collectively sent money to Turner's Lane Military Hospital or, as it was better known based on the clientele, "Stump Hospital."

Early in the history of phantom pain research, physicians thought these sensations were largely a problem with the peripheral nervous system. These theories suggested that after a limb is removed, the severed pain fibers formed a structure called a neuroma in response to the trauma. Neuromas may be unable to regulate their action potential–firing capability, sometimes sending a barrage of action potentials even without nociceptor activation. Based on this theory, an amputation farther up from the original stump could remove that clump of offending hyperactive cells, alleviating the pain. Unfortunately for these patients, they now experienced two sites of phantom pains.

The observations from repeat amputations align with a modern understanding of phantom pains. This pain likely comes from the brain. In the absence of somatosensory signals from the body after an amputation, the brain voluntarily creates anomalous pain sensations to keep itself active, maybe to prevent neuronal boredom and to minimize falling into disuse. In support of this, people born without limbs can experience phantom pains. These cases are curious and are certainly outliers. They imply that from fetal development, the default parietal lobe is wired to expect a preprogrammed set of body structures. When the body does not match up with this plan, the brain still craves those incoming signals, so it invents fictive pain to maintain traces of that default organization.

Evidence of brain plasticity supports a central reorganization hypothesis of phantom limb pain. Early-life amputations, such as those performed before six years of age, happen at a time when cortical development may be the most plastic, and these people are better able to adapt to a new body plan. There is also a clinical intervention that points toward neuronal rewiring. When a person missing a hand feels a phantom pain in one of their digits, the pain can be abated temporarily by scratching a spot on their face. Anatomically, the parietal lobe neurons that process hand and facial sensations are adjacent. Since both nature and the brain abhor a vacuum, the nearby neurons will start using those former "hand" areas and use it to process facial inputs. This invasion of territory accounts for why scratching the face can help phantom pains in unrelated body parts.

Pain, begone, I will have no more of thee! —Ira Levin, *Rosemary's Baby*

A person with neuropathic pain has the dial turned all the way up, which can be absolutely debilitating. But if this theoretical knob is turned too far to the left, that's not great either. Although far less common than excessive pain, the absence of pain is much deadlier.

In 2005, viewers of *The Oprah Winfrey Show* were given a look into that world when they met Gabby Gingras. Like any other five-year-old, she fidgets nervously on the couch, cuddling with her mother for comfort. Clear glasses wrap tightly around her face as if she just finished playing construction worker. They chitchat for a little. Then Oprah makes a joke about how Gabby would be well suited for the bitter Chicago winters, and when she laughs, you notice the conspicuous absence of teeth. Some of them broke off when she bit her toys too hard. The others were extracted to protect the inside of her cheeks, lips, and fingers, which were often bloody, never given a chance to heal. Bruises, burns, and broken bones are all a daily part of Gabby's life.

Consider another case, that of a ten-year-old Pakistani boy who gained success as a street performer. In most respects, his nervous system was normal with typical intelligence and healthy detection of warm, cool, tickle, and pressure. He just didn't experience pain. This allowed him to perform his most shocking stunt, in which he drove steel knives through his arms and then walked across a 1,000°F (500°C) bed of flaming coals. Once he sought medical attention for his shattered arm not because it hurt but because he noticed he couldn't move it. He's also missing the tip of his tongue—he bit it off when he was four.

For people with hereditary sensory and autonomic neuropathy (HSAN) like Gabby or a variation of pain insensitivity like the Pakistani street performer, everyday life can be very dangerous. Without excellent medical care and near-constant supervision, these people are always at risk of serious injury. For example, as a child, Gabby wore swim goggles all day long. Just one year before her appearance on *Oprah*, she clawed at her own eyes so severely that the left eye had to be removed and replaced with a prosthesis (her eyelids were sewn to prevent this outcome some years prior). The Pakistani street performer died from injuries sustained after jumping off a roof.

Like nearly everything in biology, pain insensitivity exists on a spectrum. In mild cases, only certain body parts are painless. With every body part affected, Gabby's type V HSAN is one of the most severe documented cases in the world. For some, the neurodegeneration and accompanying insensitivity happens gradually, first appearing in a person's twenties or thirties. Other forms are congenital. Gabby, for example, slept through the heel prick, a standard blood collection procedure performed shortly after birth. For those with less severe forms of insensitivity, the patient can have a healthy quality of life and a typical life expectancy by adopting rigorous preventive measures, such as frequent self-surveillance for skin injuries and properly fitting shoes since pain insensitivity is most pronounced in the feet.

Their sense of interoception, the ability to register internal pains such as stomach cramps and muscle aches, is also absent. An early case report of pain insensitivity described two brothers, one of whom had died following an illness that produced a fever that peaked at 109°F (42.7°C). In one case of schizophrenia-induced pain insensitivity, a physician ordered a computed axial tomography (CAT) scan right before the patient was to be discharged. They discovered a gaping hole in her upper small intestine, an injury that would cause anyone else to curl up into a ball of pure agony. Some people can even experience heart attacks without any pain.

The causes of pain insensitivity conditions are several, but one of the better-understood variations involves a mutation of SCN9A, the gene that is also implicated in allodynia. This mutation was discovered across six people from three different families, one of whom was the Pakistani street performer. All six were children born of consanguineous pairings, mostly first cousins. Each family carried a different mutation of that gene. In all cases, a single mutation resulted in a defective voltage-gated sodium channel on the nociceptor neurons. When neurons make this dysfunctional protein, the voltage-gated sodium channels stop opening when they are supposed to. Signals no longer propagate through the first-order neurons toward the spinal cord, interrupting the pain sensation at the periphery.*

In addition to those pain insensitivity conditions, a handful of psychology researchers have identified a related deficit in perception called pain asymbolia. These patients, unlike Gabby or the Pakistani street performer, can sense pain and can describe what that pain feels like. The difference is that they simply do not care about the pain. They may exhibit inappropriate responses to pain, perhaps by laughing or bringing the injured body part even closer to the noxious stimulus, the opposite of a healthy avoidance response. Just as with the pain insensitivity disorders, these people frequently have severe injuries since they too fail to protect the parts of their

* These receptors are also found on olfactory neurons, so these patients also have a loss or distortion of smell, possibly leading to eating rancid food and the inevitable painless stomach cramps.

body from being injured. Instead of having a hereditary cause, pain asymbolia is more often acquired after a brain injury, such as a stroke or head trauma. According to a 1928 German case study, the first researchers described the syndrome as follows:

> *The patient displays a striking behaviour in the presence of pain. She reacts either not at all or insufficiently to being pricked, struck with hard objects, and pinched. She never pulls her arm back energetically or with strength. She never turns the torso away or withdraws with the body as a whole. . . . Pricked on the right palm, the patient smiles joyfully, winces a little, and then says, "Oh, pain, that hurts." She laughs, and reaches the hand further toward the investigator and turns it to expose all sides. . . . The patient's expression is one of complacency. The same reaction is displayed when she is pricked in the face and stomach.*

In pain asymbolia, the source of pathology is likely in the brain rather than in the periphery. Asymbolics experience the entire nociceptive chain of signaling in the periphery and part of the central nervous system. Noxious stimuli on their skin still cause the opening of the $Na_v1.7$ channels and neuronal depolarization. These neurons still form synapses onto the second-order neurons in the spinal cord, and these cells still release neurotransmitter into the thalamus. But somewhere in the brain, there is a disconnect between that noxious signal and the perception of pain. Maybe these people no longer have the appropriate circuitry that ties pain with an emotional response. Or maybe they no longer understand why that stabbing sensation might lead to injury. Regardless of why, they are unmoved in the face of pain.

While cellular changes provide a strong mechanistic explanation for HSAN, pain asymbolia is not as well understood. Brain-imaging scans of asymbolics suggest that the one structure that is often damaged is the insular cortex. The insula has strong reciprocal connections with the amygdala, implying its involvement with the processing of emotions. The insula also receives inputs from the thalamus, which is where the third-order pain neurons reside. In theory, when the insula gets damaged by a stroke, the noxious sensory inputs through the thalamus do not successfully pass through the insula to tell the amygdala to initiate a fearful response to the injury. One theory suggests that pain asymbolia presents akin to depersonalization, when a person loses the sense of "ownness" over their emotions or actions, as if they are watching someone else experience those things. However, in the case of asymbolia, only pain is depersonalized, while other emotional processing remains intact. If true, it provides further support that the locus of this condition is central rather than peripheral.

> "Wretch," I cried, "thy God hath lent thee—by these angels he hath sent thee Respite—respite and nepenthe from thy memories of Lenore; Quaff, oh quaff this kind nepenthe and forget this lost Lenore!" Quoth the Raven "Nevermore." —Edgar Allan Poe, *The Raven*

Balance is essential in all things, and pain is no exception. Too much, and daily life is hell. Too little, and life is short. There are instances, however, when someone would want to purposefully tip this balance one way or another.

Because a defective Na_v1.7 protein leads to both excess pain and pain insensitivity, it is no surprise that these and other voltage-gated sodium channels are molecular targets for local anesthesia. The best-known example of one such inhibitor is lidocaine, the substance injected into your gums during a trip to the dentist.* Before the adoption of lidocaine, there was novocaine, a drug with

* Dental pain is a common fear and has been depicted in various media, including Edgar Allan Poe's *Berenice*, a dark (even by Poe standards) tale of tooth pulling after a live burial; the comedic song "Dentist!" from the quirky off-Broadway *Little Shop of Horrors*; and the brutal tooth-extraction-by-claw-hammer scene in the Korean revenge tale *Oldboy*.

a similar mechanism of action. And before novocaine was another "-caine" drug, the original voltage-gated sodium channel inhibitor used as a local anesthetic for dental work: cocaine. And to absolutely no one's surprise, one of the first people to hint at the anesthetic application for cocaine on the gums was Sigmund Freud.*

Local anesthetics are useful under certain circumstances, but sometimes you need something more potent, something to completely quell the pain like nepenthe, Homer's literary invention. This is the role filled by opioids, chemicals that bear some resemblance to opium. These drugs, which include heroin, morphine, oxycodone, and fentanyl, to name only a few, decrease cellular activity when they bind to and activate their corresponding opioid receptors. The location of these opioid receptors provides some clues about how they exert their effects on a person.

Opioid receptors are found densely in the PAG. Neurons in the PAG come in at least two different flavors. One population, the ones that make up the descending pain modulation pathway, sends their projections down toward the raphe nucleus. They receive signals from another population of neurons, interneurons that are continuously providing background inhibition of those projection neurons. Between these two cell types, the inhibitory opioid receptors are on only the interneurons, not the output neurons. So when a person takes an opioid, it decreases the activity of inhibitory neurons, which increases the activity of the PAG and hence the descending pain modulation system. Since opioids also disinhibit the dopamine circuits in the VTA, they can produce a pleasant euphoric effect at doses at and above those recommended for pain relief. This increase in the dopamine signal through the mesolimbic pathway encourages further opioid use.

* Medical uses of cocaine, unlike Freudian psychoanalysis, haven't completely fallen out of favor today. Preparations of cocaine are still used today for surgeries of the face, particularly around the nasal cavity. Not only does it block the pain inputs, but, as a bonus, it acts as a blood vessel constrictor, making for a cleaner operation site.

The combination of increases in PAG and VTA activity contributes to the so-called runner's high, the feeling you get after reaching a certain point in a strenuous workout. Compounds released by the hypothalamus are pumped into the blood through the pituitary gland. One such compound is beta-endorphin, a name that is a combination of "endogenous" and the suffix "-orph," as in "morphine," so called because they activate opioid receptors. That little bit of pain blockage and burst of feeling good might be just enough to push us through to the finish when the going gets tough. To add an extra layer of complexity to the story of opioid action, receptors are found very densely in the anterior cingulate, the cortical areas that correlate physical sensations with emotional content. When opioids change these circuits, they block the formation of the emotional content of pain, akin to inducing a temporary state of pain asymbolia. Between the triple mechanisms of nociceptive neural inhibition through the descending pain modulation system, the enhanced feel-good sensations through midbrain dopamine, and emotional memory blunting through anterior cingulate modification, it's no wonder that opioids are so effective at doing their job.

Although most pain research is focused on lowering pain, there has been a tiny bit of work done on those rare cases when you want more pain, such as to help people with HSAN or pain insensitivity. One possible target for treatment is to focus on the endogenous opioid system. In addition to beta-endorphin, our brains make a host of neuropeptides that act at opioid receptors, one of them named enkephalin. These signaling molecules are increased in cases when the levels of $Na_V1.7$ protein are decreased. So less signaling upward and more opioid receptor activation in the brain block pain on two fronts. Antagonists such as naloxone may be helpful in restoring the healthy, protective sense of pain.

No tears, please. It's a waste of good suffering. —Clive Barker, *Hellbound Heart*

CHAPTER SIX

REMEMBER TO FORGET

MEMORY AND AMNESIA

I had hundreds of megabytes stashed in my head. —William Gibson, *Johnny Mnemonic* (1981)

How much am I carrying? 320 gigabytes. —*Johnny Mnemonic*, film adaptation (1995)

IN 1981, EXTERNAL STORAGE MEDIA HELD HUNDREDS OF KILObytes per floppy disk, and commercially available home computers rarely had hard drive capacities of more than a few megabytes. So when William Gibson published his cyberpunk science fiction short story about a data courier hired to transmit sensitive information, the idea of hundreds of megabytes was pretty spectacular although not outside the realm of imagination. Hundreds of megabytes! Amazing! Of course, when Keanu Reeves played that same data courier in the 1995 film adaptation of *Johnny Mnemonic*, the dialogue had to be adjusted with appropriately scaled-up numbers to accommodate the advancement of technology. People's hard drives at the time maybe held a few gigabytes of space, so hundreds were an impressive number. If *Johnny Mnemonic* were to be remade as of the time of writing, the data courier might brag about the hundreds of terabytes of valuable corporate data stored somewhere in his head.

According to a major framework of how memory works, there are three steps to the process. First, the incoming information must get encoded, the process of translating sensory signals into a pathway that the brain can use later. Then that information must be stored, tucked away in circuits that can persist over the course of a period between seconds and years. Finally, that sensory experience has to get recalled, which brings back the sensory experience. Failures at any point along this process might manifest in different ways: forgetting someone's name within seconds after an introduction is made (encoding), misplacing your car keys and wallet (storage), or brain farts (recall).

This process can be enhanced by mnemonic devices.* The memory palace is one of the oldest of such tools, described by the ancient Greeks. According to legend, a man stepped outside during a banquet when the building collapsed. As the sole witness to the tragedy, he was asked to identify the bodies based on who was sitting where. His successful mental reconstruction of the scene led him to develop his memory palace technique. Here, a person imagines a physical space in their mind, perhaps a childhood home. During encoding, they mentally walk through that space and imagine the sensory inputs in different parts of the rooms. The more items they are tasked with remembering, the more rooms or spaces they use. Then, when they are recalling those memories, they retrace their steps throughout the palace, imagining what they had put in each part of that space. For example, they might picture themselves outside of a house where they see two big bushes, seven paved stepping stones leading to

* A couple of notorious phrases have helped countless students of neuroscience memorize the names of the 12 cranial nerves and the type of information carried by them. Although modern adaptations of these phrases have substituted the lewd words for ones more family appropriate, many neuroscientists of a certain age can't seem to un-encode sayings about things that people do under the sheets or advice about which traits to look for in a marriage partner.

the front door, and a glass window divided into four sections.* The idea is to pair an abstract idea (the number seven) with something tangible and physical (moving your legs seven times while seeing seven pieces of rock), thereby engaging other brain functions and providing context to the number.

Memory formation is enhanced by previous experience and improves when new memories incorporate old memories, like using an existing scaffold to build a new construction. A group of chess players were shown one of two chessboards with the instruction to memorize the locations of the pieces. The experimental condition was a setup from an actual game with pieces in positions that would appear in realistic play: kings tucked away in their corners while the other pieces fight for dominance over the center of the board. In the control condition, the same pieces were placed on the board at random, sometimes in unlikely or otherwise impractical positions, such as kings exposed to threat of check, doubled pawns, and even two bishops on same-colored squares. When asked to recall piece position in the genuine play condition, there was a near-linear correlation between how many pieces were remembered and that player's skill at chess as quantified by the international Elo rating system. The conclusion here suggested that players with more experience use their extensive knowledge of the game to aid in memory. "Oh, this board state looks like one that you would see in the Queen's Gambit declined," a player with moderate knowledge of chess openings might think. "Gee, this position reminds me of when Magnus Carlsen won the final game of the 2021 World Chess Championship. It's mate in six," a master of the game would observe. Chess skill, however, did not matter for memory in the control condition since even an encyclopedic

* Reciting the sequence 2-7-4 would break a number sequence memorization record set by Akira Haraguchi in 2015. However, you'd also have to remember the preceding 111,699 digits of pi.

knowledge of chess openings or historic matches wouldn't help players memorize pieces placed at random.

Another observation of note is that memory is selective. For most people, it is impossible to encode everything. Our sense organs are constantly taking in massive amounts of information all the time, including hundreds of objects in our visual field at any moment, all kinds of background noises, smells, internal states such as hunger and thirst, and bodily tactile sensations. We're not even consciously aware of most of that incoming information.* Instead, only the most salient stimuli that receive attention get processed, akin to how we can isolate our friend's voice at a noisy party while everything else fades into the background. This difference in attention is one reason why reading a romance novel by the pool and studying from a textbook in the library are two distinct actions. Yes, both involve using your eyes to look at rows of words from left to right, over and over again. Leisure reading just uses fewer attentional mechanisms. In contrast, reading intensely into the deeper implications of sentences and thinking about how this knowledge might be assessed while preparing for a test engages deeper attentional mechanisms that improve encoding.

For a memory to be maintained over time after encoding, it has to be "put" somewhere. The leading hypothesis about precisely where any given memory is held is the engram, a theory proposed more than a century ago. The basic idea is that once a sensory stimulus receives the right attentional process, encoding of that stimulus requires that some pattern of cells gets changed in some way. This change must be persistent enough that it remains even after that stimulus is no longer present. Then, at some point later, reactivation of the cells in the engram brings that particular memory back to mind. The engram itself is a vague entity, in part because

* For some anecdotal evidence in support of this, I promise that you weren't thinking about your posture, the feeling of fabric on your shoulders, or how deeply you are breathing—until right now.

it has proven to be a huge challenge to definitively demonstrate its existence in a laboratory setting. Looking back on the early descriptions of the engram, even the creator of the term wrote in 1923 that understanding the molecular parts of the neural trace was a "hopeless undertaking." But as molecular biology expanded its tool kit over the course of the century, more and more details emerged over what the engram might actually look like.

This is where plasticity enters the discourse.

Of the buzzwords in neuroscience, few are as broadly defined as plasticity. In general, the word is used to refer to the ability for the nervous system to change. On the cellular level, plasticity might look like neurons that physically change shape, such as when dendrites, the receptive sites of the neuron, get larger, wider, or longer or disappear completely. It might be when an axon extends a new branch, bridging a communication route between cells that were once isolated. Plasticity also exists at the level of molecules. Neurotransmitters can temporarily weaken a communication pathway by clamping down on the release of another neurotransmitter. Modifications to the structure of receptor proteins can make them open quicker and close slower, meaning that the same activation of that receptor now sends a bigger signal. Some plasticity mechanisms can turn genes on or off.

One of the earliest laboratory demonstrations of plasticity was performed on funny little sluglike gastropods called the California sea hare (*Aplysia californica*). These ocean-dwelling critters are often found inching across the tidal plain of warm tropical waters. Despite being about a foot long, these creatures have relatively rudimentary nervous systems made up of only 20,000 neurons. The connectivity of some of their nerve cells are so well characterized that many *Aplysia* neurons have been given specific alphanumeric names, such as B51 or R2.

And yet, even with such a simple nervous system, *Aplysia* can form memories, one of the most complex behaviors in all of neuropsychology. The experiments demonstrating learning in *Aplysia* were conducted by Dr. Eric Kandel based out of New York University's medical school. Kandel's team examined a behavior that *Aplysia* exhibit called the gill withdrawal reflex. This is a two-organ reflex circuit where a sensation on its tough, tubelike siphon leads to activation of motor neurons that retract the gills. In theory, this behavior is an ancient evolutionary response, protecting *Aplysia* from whatever dangerous predators are stalking around in the water. In Kandel's laboratory, a paintbrush stroked gently across the siphon initiated the gill withdrawal reflex. After this innocuous stimulus was repeated, the strength of the gill retraction became progressively weaker, a demonstration that the animal learned that there is no danger associated with the brush.

Another goal of these experiments was to figure out the neural circuitry involved in the reflex, and this is where the advantages of working with *Aplysia* really came to the forefront. Some of their neurons are huge in comparison to mammalian neurons, such as the gargantuan R2 neuron, which is about a millimeter in diameter. So when it came time to study individual neurons, their large size simplified the task of penetrating the neuron using a tiny glass electrode. One by one, Kandel systematically probed neurons, delivering tiny electrical pulses while watching and waiting for the gill response. Then one fateful 1968 afternoon, he came across cell L7. This neuron, when activated, caused a robust retraction of the gill even without touching the siphon. He had found the shortcut to the reflex arc. By the end of this meticulous mapping, he discovered a total of six neurons that controlled the gill.

Now, to tie the whole story together, Kandel wanted to demonstrate that behavioral learning is a result of changes at the cellular level. The electrodes used to puncture the neurons could also be used to observe subtle changes in the electrical properties of the neurons. On first touching the siphon, the motor neuron responded very strongly. After the animal became desensitized to the sensation over repeat brushstrokes, the motor neuron response steadily declined. This observation was—and still is—the ideal outcome for many learning studies in neurobiology: finding a quantifiable behavior in the intact animal and then finding the neurons that are responsible for that behavior, from animal to cell.*

The dampening of reflexes in *Aplysia* is an example of synaptic depression, a form of plasticity where a connection weakens. But a synapse can also get stronger with experience. The most well-studied phenomena of synaptic strengthening is long-term potentiation (LTP). One of the original demonstrations of LTP was performed in the hippocampus of anesthetized rabbits.† Here, two researchers at the University of Oslo inserted thin tungsten wires adjacent to the perforant pathway, one of the main input structures that feeds signals into the hippocampus. Then a glass electrode such as that used by Kandel in his studies was placed in the dentate gyrus, a section of the hippocampus downstream of the perforant pathway. By delivering a small pulse through the perforant pathway, a bit of excitation could be picked up by the electrode, demonstrating a connection between these two areas. Then, by delivering a high-intensity stimulation train (100 pulses per second for three seconds), the detected signal was increased, sometimes enhanced threefold. Most spectacularly was the time frame of this increase. In some rabbits, the increased synaptic strength was still observed 10 hours later. Later experiments with a similar design have found that hippocampal LTP can persist for up to a year. This plasticity probably persists for decades, considering that we still hold memories from first and second grades.

* Before this work, Kandel fled Austria to escape the horrors of Nazism. After this work, he was awarded a Nobel Prize in 2000. Today, he is best known by neuroscientists as the author of *Principles of Neural Science*, a massive tome of more than 1,600 pages that is the gold standard for a graduate education in neurobiology.

† Neuroscience research involving bunnies has been decreasing steadily since the early 1990s. Today, rabbits are used in the study of infectious disease and veterinary medicine and occasionally as props in Jordan Peele's *Us*.

The mechanism behind LTP can be seen at the synaptic level during and immediately after that high-intensity stimulation train. The perforant pathway contains axonal fibers that release the excitatory neurotransmitter glutamate. Once released into the synapse, those molecules of glutamate can bind with receptors found on the postsynaptic cell, such as the AMPA and NMDA glutamate receptors. While both are excitatory, the NMDA receptor does not get activated unless there is already some degree of excitation affecting the cell. NMDA receptors also allow calcium into the neuron, where it functions as a signaling molecule, binding with and turning on certain enzymes. One such enzyme brings more AMPA receptors to the postsynaptic membrane. So when a high-frequency stimulation train is delivered, massive amounts of glutamate are spilled into the synapse, which activates both AMPA and NMDA receptors, allowing calcium through their gates and in turn activating enzymes that bring more AMPA receptors to the surface. The next time a little bit of stimulation releases some glutamate, there are more AMPA receptors available for binding. Hence, the same amount of glutamate release causes more excitation.

This mechanism agrees with theories about the engram. Imagine there is a circuit that encodes a tidbit of information. Each time that circuit is activated, there is a possibility that a bit more glutamate is released, which could activate both AMPA and NMDA receptors sufficiently to induce that trafficking of receptors to the surface in a positive feedback loop. This cascade of activity has been summarized by Donald Hebb in the catchphrase "Cells that fire together, wire together."

Yes, LTP is probably the way we learn things, and learning new things is usually good. Successful protohumans learned the faces of caring family members. They learned which berries were safe to eat. They learned to avoid growling animals with pointy teeth. But under some circumstances, LTP can also be bad. It's prudent to keep in mind that potentiation is just a tool, one of many that the nervous system has. Consider how a fireman would use an ax to break open a locked door to rescue a family from a burning building but how *The Shining*'s Jack Torrance might put that same ax through a door for a totally different purpose. Consider the following examples:

- Exposure to a single dose of cocaine causes a potentiation of the excitatory synapses onto the dopamine neurons of the ventral tegmental area, the cells that start firing more often when exposed to cues associated with "something good," such as more cocaine. The implication here is that plasticity primes us for the next exposure to that "something good," possibly suggesting why some recreational cocaine users become compulsive users.
- Stressing out a rat, perhaps by stranding him atop a tiny plastic island surrounded by a pool of water or by stuffing him into a narrow plexiglass tube, causes LTP in the basolateral amygdala. These enhanced excitatory inputs lead to a hypersensitive amygdala that initiates the fear-related physiological cascades at a lower stimulus threshold. Like Jigsaw torturing his victims, psychological stressors induce a similar LTP: social defeat, where an experimental rat is placed in a cage with a bigger, more aggressive bully rat, can lead to a reshaping of amygdala neurons, now with more branches to receive inputs.
- In the healthy spinal cord, neurons of laminae I and II receive much of the axonal inputs from pain-sensing organs. Here, some neurons project upward into the brain, carrying the message of pain. After an injury, the spinal cord exhibits an amazing degree of plasticity, where non-pain somatosensory neurons physically sprout into laminae I and II, causing someone to feel pain when it doesn't genuinely exist.

Your wife always used to say you'd be late for your own funeral. Remember that? Her little joke because you were such a slob—always late, always forgetting stuff, even before the incident.
—Jonathan Nolan, *Memento Mori**

Memory is like a bridge. It ties our past experiences with present decisions that affect our future lives. Without this bridge, we get stuck in the present.

Since his brain injury, *Memento Mori*'s protagonist Earl can no longer create new memories. His room is plastered with signs reminding him to go back to sleep if it's still early, to brush his teeth after waking, and to check for lit cigarettes before lighting a new one. He ignores the last sign only to discover a lit cigarette already burning in his ashtray. He uses clues found on these signs and the occasional photograph to piece together his own history. His body is covered in tattoos, a permanent reminder of his sole objective in his fractured life: revenge.

Earl and his amnesia would be the perfect case study for Dr. Brenda Milner. Regarded as the founder of neuropsychology, Milner is one of the most impactful researchers in all of neuroscience. From the beginning of her PhD research, Milner was interested in high-order intellectual processes. She established herself in the field of experimental psychology with her 1954 article "Intellectual Function of the Temporal Lobes." Here, she describes the variety of deficits that manifest after temporal lobe injury, including deficits in visual and body sensation and a loss of speech and language.

She would soon revolutionize the way we think about memory altogether. Her most groundbreaking work was a series of case studies about a man, Patient HM. Born in 1926, HM suffered

* This piece was published in *Esquire* magazine, but a related story of a clinically accurate amnesic man out for revenge was also told by his older brother Christopher Nolan in his 2000 crime thriller *Memento*. Apparently, Jonathan found his inspiration in an undergraduate psychology class.

from childhood seizures brought on by an unlucky convergence of a genetic predisposition and a head injury sustained after falling off his bicycle. At 10 years old, the seizures started small. In an absence, or petit mal, seizure, HM would suddenly stop speaking mid-sentence, stare blankly out into space, or perform subtle but repetitive actions, such as lip smacking or chewing. Instead of outgrowing the seizures as some children do, HM's only got more severe. At 15 years old, he experienced his first generalized tonic-clonic, or grand mal, seizure, the kind that prompted an immediate visit to the hospital. These are the kinds when the patient loses consciousness and control over their body as their muscles tense up. They might thrash their arms rhythmically or bite their tongue. Despite being prescribed a massive cocktail of antiepileptic drugs, HM's frequent convulsions were still uncontrollable. His attacks forced him to drop out of high school, and at 27 years old, HM was forced to quit his job at the typewriter assembly plant. Seizures were destroying his life.

In search of a new approach to treatment, HM's family brought him to the neurosurgeon William Beecher Scoville in 1953. Somewhere within HM's temporal lobe, there was a group of cells that behaved differently from the others, perhaps firing more action potentials than expected. Once these cells started firing, their aberrant activity spread across the network of connected neurons like a row of lined up dominoes, producing a chain reaction that led to the seizures. Knowing this, Scoville proposed a "frankly experimental operation": complete removal of the temporal lobes from both halves of the brain. HM agreed, desperate to regain some normalcy in a life rent asunder by medical emergencies and hospital visits.

HM was prepped for surgery. Scoville drilled two holes in the skull, each one and a half inches in diameter, just above the eyes.*

* This step of the operation isn't without precedent. Archaeologists have unearthed human skulls with fully healed head holes dating to around 10,000 years ago.

Then an electrode the size of a postage stamp was inserted through the new holes and placed against the surface of the medial temporal lobe to better narrow down which patch of cells was responsible for the seizures. Not seeing a clear locus of activity, Scoville went forward with the extraction of bilateral medial temporal lobe structures. The hippocampus was almost completely excised. The amygdala was likewise resected. The parahippocampal gyrus, the entorhinal cortex, the uncus—all of it scooped out.

By some measures, Scoville's procedure was successful. HM no longer experienced the life-threatening seizures that had once interfered with his day-to-day activities.* But there was a side effect that became apparent soon after his recovery. HM developed a severe case of amnesia. He was unable to recall anything that happened to him in the years after the surgery. He couldn't remember the names of celebrities who appeared daily on his television or the public figures who were discussed in every newspaper headline. He couldn't remember any instance of being brought into the hospital for his frequent health examinations. He couldn't remember flying to Montreal to visit Dr. Brenda Milner, one of the many psychiatrists who examined HM's unique brain.

He also never remembered failing a barrage of memory-related neuropsychiatric tasks. For example, he did not form new memories of words that entered the English language since his 1953 surgery. He identified old words ("butcher," "shepherd") as real and made-up words ("thweise," "phlawse") as fake but was practically coin flipping on those newfangled words, such as "hacker" and "psychedelic." On an association task, he could not remember pairs

* In the years of medical progress, modern approaches to treating epilepsy have improved. Today's seizure-controlling drugs have fewer adverse side effects. One fancier treatment option is to implant an indwelling medical device that detects changes in the nervous system that precede a seizure. When it detects that signal, it responds by stimulating the parasympathetic system, which signals the body's relaxation response. As a last resort, however, removal of brain tissue is still performed although with moderate success: Slightly more than half of these patients are seizure free following recovery.

of words when one of the two was presented to him even if they had some shared context, such as "rose" and "flower." Despite being a consummate TV watcher, he couldn't remember the names of the current presidents, from Kennedy through Nixon. When asked how old he was or to approximate the current year, it seemed like he was guessing. Sometimes his answer would be off by as much as a decade. This skewed perception of time was succinctly summarized by one of Milner's students, Dr. Suzanne Corkin, when she titled her biography of HM *Permanent Present Tense*. It should be noted that he actually scored slightly above average on an IQ test years after his surgery, showing that his deficit was not of generalized intellect but specific to memory formation.

He didn't fail at every memory test, however. When interviewed about his childhood memories, many were still intact post-surgery. HM remembered how the stock market crash of 1929 had affected his family growing up. He remembered car trips with his family and could recognize the location of old photographs of himself on vacation, sometimes even telling stories about the pictures. He could even describe specific memorable events in accurate detail, such as when he, as a 13-year-old, was allowed to pilot a single-engine airplane through the sky. But all the events of his life after his surgery were lost to time.

There were other tests, such as the very challenging mirror-drawing task, at which HM was quite adept. Imagine two stars in outline, one slightly smaller and completely contained within the larger one. His task was to draw a third star in the narrow space between the two stars without crossing any of the outlines that serve as a boundary. The difficult part, as the name implies, is that he must complete this task while watching his hand in a mirror, where movement is reversed. Over the course of 30 practice runs spread across three days, he made fewer mistakes and completed the drawing faster and faster, improving from 30 seconds on his first attempt to a very impressive three seconds after repeat training. As expected, based on his inability to create new memories about events in his life, HM had no recollection of ever drawing any stars in the mirror. Each day of testing, the researchers would introduce themselves to him, explain to him why he was seated in front of a mirror, and describe to him what his objective was. Yet it is apparent that he learned something through all that training. At the end of the final drawing session, HM even remarked, "Well, this is strange. I thought that that would be difficult, but it seems as though I've done it quite well."

Because only certain aspects of HM's memory were lost after his surgery while others were maintained, Milner concluded that a more accurate model of memory is not a single construct. Some types of memories require the hippocampus and medial temporal lobe structures to function, such as events that happened to a person or new facts and pieces of information. The former came to be called episodic, or autobiographical, memory, while the latter was called semantic memory. Both episodic and semantic memories fall under the umbrella of explicit, or declarative, memories. Milner reasoned that because HM lost this form of memory after his lobectomy, these memories must be processed by circuits that pass through these missing brain structures. On the other hand, motor coordination tasks, such as the mirror-drawing test, is an example of an implicit memory. HM improved on the mirror-drawing task, and his ability to learn this skill indicates that implicit memories are processed independently of hippocampal circuits. Implicit memories are unconscious memories in the same way that someone proficient at riding a bicycle can coordinate many major muscle groups to balance and move themselves forward without much conscious, active thought.

HM also taught us some interesting tidbits about the time frame of memory loss. The terms "anterograde amnesia" and "retrograde amnesia" are used to describe when the amnesia begins relative to an inciting event. HM's memory loss was strictly anterograde, as he couldn't make new memories after the surgery, while his childhood memories were still intact. In people with retrograde amnesia, they lose memories prior to the incident.* In the context of neuroanatomy, the conclusion is that the hippocampus plays a major role in the encoding process and less so for storage or recall.

HM is widely regarded as the "yardstick" by which other cases of amnesia are measured. But there have been many other highly instructive studies of individuals with memory impairments, each of which teaches us something interesting about other facets of amnesia:

- A retired postal worker, Patient RB, experienced anterograde amnesia of explicit memory after surviving a stroke. Postmortem analyses found his brain to be mostly intact except for a tiny region of the hippocampus called CA1, which was almost completely destroyed. Study of RB's deficits added evidence that the hippocampus, not the many other structures removed in HM's surgery, is the most critical component for processing explicit memories. Furthermore, it demonstrated that signals through the hippocampus happen in a stepwise sequence, and when one of those links in the chain falls apart, the healthy parts are unable to compensate, and the entire structure fails.
- In a demonstration that amnesia severity exists on a spectrum, British musician Clive Wearing has a short-term memory of just a few seconds. He frequently loses his train of thought by the time he gets to the end of his sentences. He will forget who is in the house with him, greeting his wife joyfully every time she reenters the room. Despite the severe deficits in his explicit memory, his implicit memory is

* "Hollywood" amnesia is often shown as retrograde, where the person cannot remember some details about their past. Genuine clinical cases of amnesia are more *Memento* and less *The Bourne Identity*.

well preserved. An accomplished musician, Wearing has no difficulty conducting a choir or sight-reading sheet music.

- Patient NA taught us that the processing of verbal memories relies much more on circuits around the thalamus than on the hippocampus. His language-specific amnesia appeared after a toy sword pierced his brain through his nasal cavity.

For the sake of maintaining patient confidentiality, HM remained anonymous until his death in 2008. His name was Henry Molaison. He lived to be 82 years old.*

Clementine Kruczynski has had Joel Barish erased from her memory. Please never mention their relationship to her again. Thank you. LACUNA, LTD." —*Eternal Sunshine of the Spotless Mind*

During the procedure, Joel Barish changes his mind. One by one, memories of his ex-girlfriend get zapped away. She already erased all her memories of him. Sure, their relationship ended poorly, but replaying their moments together as they are slowly eliminated, he is reminded that there was overall more good than bad.

Forgetting is the counterpart of memory. Considering that by far we forget more things than we remember, forgetting might actually be the default action. Forgetting happens when some memory that has been encoded is not recalled successfully. Maybe among the engram that stores a particular memory, not enough neurons are recruited during recall, sending a diluted signal that we cannot piece together into a complete memory. Maybe non-engram neurons are also activated during recall, resulting in interfer-

* Molaison was well-liked by Milner and her research team. She lamented his loss, feeling as though she had lost a friend—a friend who never remembered her. He was such an important man in the world of neuroscience that I remember exactly which classroom I was in and the professor who told us that he had passed.

ence from other memories and causing past events to be blended together. Or maybe those plasticity changes between specific neurons made during encoding get reversed, causing the engram to collapse completely.

Forgetting is not just an incidental process. Animals actively expend energy, producing biochemical factors that purposefully weaken engrams. The fact that our precious cellular energy is sometimes spent on the loss of specific memories suggests that forgetting may be adaptive. Maintaining a bunch of facts is helpful during test taking or pub trivia, but neither of these things was helpful to the animals of early evolutionary time. Instead, these might clutter up the brain, slowing down processing of stimuli that threaten survival. When faced with a nine-foot-tall bear, little details like ear pointiness or head roundness are irrelevant. Why bother sorting through all the mental records documenting the aggressiveness profiles of different types of bears? It's much easier to just maximize survival odds by simply avoiding them altogether.

Forgetting also helps our brain be more flexible, which is one of our best strengths. Sometimes, we may hold a certain expectation about the world, but then we learn something new that contradicts previous knowledge. When reality doesn't align with our past experiences, new learning requires that we rewrite that engram. Being flexible likely allowed early hominids to outperform other animals: tool use requires improvisation, and so does wayfinding and navigation if a usual route is blocked by, say, a bear. Implicit memories especially benefit from flexibility, where the occasional forgetting offers a boost to creativity by keeping us from doing the same action over and over again. For example, imagine food-gathering strategies never changed even though the climates and seasons did. A diet of nuts and berries doesn't fuel our expensive bodies as well as fish and bone marrow.

Most complex of all, there are prosocial benefits to forgetting. People change, and without forgetting, our mental image of everyone is still at the age we last built memories with that person. There are also mental health advantages to forgetting some of our autobiographical memories. Many of us (thankfully) cannot remember those cringe moments from junior high, embarrassing party fouls of our twenties, or social faux pas of adulthood. "Blessed are the forgetful," as Nietzsche says.

However, here's the thing about forgetting. You don't get to pick which memories fade away. If a person, like Joel, decides they want to speed up this process, they might consider a procedure that deletes memories, the promise offered by Lacuna, Inc., of *Eternal Sunshine*. As with most science fiction storytelling, the specifics aren't explained, nor does it matter. Memory deletion is already possible today. Consider these options for amnesia therapy:

- You could try one of the most potent amnestic drugs in existence: alcohol. Short-term acute intoxication produces spotty anterograde amnesia for the duration of the drug effect, but to delete memories from farther back in time, the drug would have to be in the system at exceedingly high doses for a prolonged period. Long-term alcohol users are at risk for developing Korsakoff syndrome, diagnosed by complete retrograde and anterograde amnesia for autobiographical memories. Alcohol interferes with the body's processing of thiamine, a vitamin essential for ATP production. Low ATP leads to cell shrinkage or death, which accounts for the atrophy of the brain in patients with Korsakoff. Sure, this strategy will likely wipe any traces of Clementine from Joel's memory, but other memories are also subject to the amnestic effects of this brain damage. In addition, since the neuron loss is not specific, there would be other undesired side effects, including dysfunction in the frontal cortex and cerebellum, leading to personality changes, poor impulse control, and difficulty with balance.

- Milder amnestic drugs produce memory loss without the severe brain damage of Korsakoff. Many of these drugs work by increasing the action of the neurotransmitter GABA, the main inhibitory signal in the central nervous system. They are typically prescribed for a range of psychiatric conditions, including epilepsy, anxiety, insomnia, and panic attacks. During memory formation, the steps that strengthen connections within the engram require the activation of neurons. Decreasing neuronal activity therefore makes it less likely that the engram is successfully maintained, which may explain why memory loss is a common side effect of these drugs. The amnesia brought on by these meds is more anterograde, which wouldn't be helpful for the heartbroken Joel.
- Head trauma sometimes produces retrograde amnesia. After an injury, the body reacts by spilling a cocktail of inflammatory molecules into the bloodstream. Normally, these molecules signal to the immune system where an injury is located and how to repair the damage. But in a head injury, these signals flood into the brain. These signals might disrupt the neurons that make up the Clementine engram, therefore producing the desired retrograde amnesia. This amnesia is often temporally graded, with recent memories being the most vulnerable to deletion, so Joel's childhood memories will likely remain intact. The downsides are obvious, however. The most likely outcome of head injury is bleeding in the brain, which is always bad.

Obviously, these strategies would lead to overwhelming masses of dissatisfied customers.

Considering the available options and their shortcomings, the criteria for a successful procedure become clear. The amnesia produced by these strategies is like a sledgehammer in the hands of an amateur, recklessly demolishing memories with wide-reaching consequences. In a successful procedure, the amnesia must be absolutely specific both in time frame and in the memories targeted. Ability to create new memories must remain intact, and the memory loss should target only Clementine and none of the other memories formed at that same time. Targeted deletion of specific memories is still much in the domain of science fiction, at least in nonhuman research. So could it be possible to identify Clementine-specific neurons then mark that engram for deletion?

I present here "An Unethical Guide to Erasing Joel Barish's Memories of Clementine Kruczynski, Based on Speculative Biotechnology That Has Never Been Tested in Humans."

The first step is to figure out which neurons are responsible for the engram that stores Clementine memories. As difficult as this might sound, work like this has been done and, surprisingly, in people. The landmark study was a 2005 paper describing single-neuron recordings of cells in the medial temporal lobe from awake patients undergoing epilepsy surgery. A dense array of electrodes is pressed into the brain tissue, which can detect the action potential firing of individual neurons. Then the researchers showed the patient different images. One neuron in particular was silent at rest. It was silent in response to pictures of the Sydney Opera House, the Tower of Pisa, or the Golden Gate bridge. It was silent in response to pictures of animals, such as snakes, spiders, or dolphins. It was silent in response to pictures of famous people, including Kobe Bryant, Bill Clinton, Pamela Anderson, or Julia Roberts.

But when a picture of Jennifer Aniston was shown, the neuron came to life with activity. The angle, the lighting conditions, the hair color or style, the facial expression: Differences in these parameters don't seem to matter. If it was Jen, this neuron lit up.

These neurons came to be called, appropriately, Jennifer Aniston neurons. Using a similar approach, Halle Berry neurons were also identified. These neurons started firing in response to images of Halle Berry, screen captures of her wearing her iconic black rubber mask from her role in *Catwoman*, a black-and-white drawing of Halle Berry, and even just the words "Halle Berry." Also identified were neurons for groups of people and fictional people, such as the Beatles neurons, *The Simpsons* neurons, and Luke Skywalker neurons, each one firing with tremendous specificity to one stimulus and nothing else.

The stimulus-dependent increase in neuronal firing is a key element in this process since increased activity is also a common feature of LTP. When neurons increase their firing rate, a class of genes called immediate early genes (IEGs) gets turned on. Once these IEGs get activated, so does the machinery involved in LTP, processes that increase receptor expression. In mouse experiments, the genetic sequence of their neurons can be changed to code for a fluorescent reporter molecule under control of the IEG promoter. Now, whenever that cell is activated, the IEG turns on both the plasticity mechanisms and the reporter gene, causing only the highly active cells to glow green.

Next is to replace that reporter molecule gene sequence with a different construct, one that will allow for the reversal of LTP. This will have to rely on a technique called chromophore-assisted light inactivation. This strategy codes for a molecule that goes by the ominous name KillerRed. When struck with green light, KillerRed causes the destruction of other proteins, notably one called actin. Actin gives structure to cells, roughly analogous to the neuron's skeleton. Actin is especially critical for the maintenance of dendritic spines, one of the neuronal features that changes during LTP. By breaking down the structures involved in learning, KillerRed reverses plasticity, hopefully without destroying the neuron altogether.

For Joel, all he would have to do is bring in all the possessions that remind him of Clem. Once ready for the procedure, it begins with a tiny intracranial injection of a harmless virus that will insert the gene construct of KillerRed under the control of a specific IEG promoter, say, c-Fos or arc. In a sensory deprivation setting, the items would be presented to him along with images of Clem, which would increase activity of the Clementine Kruczynski neurons and their related engram no matter how widely distributed across the cortex they are found. Activation of these neurons will cause the KillerRed protein to be made in the neurons. When it's deletion time, shining light directly into the cortex will initiate the cascade of molecular destruction that results in the breakdown of dendritic spines, erasing the engram and any memory traces of this aspect of his past.

I propose using a flash of light to delete a memory. Let's call it the *Men in Black* treatment.*

> Programming an artificial memory of a trip to another planet—with or without the added fillip of being a secret agent—showed up on the firm's work-schedule with monotonous regularity. In one month, he calculated wryly, we must do twenty of these . . . ersatz interplanetary travel has become our bread and butter.
> —Philip K. Dick, *We Can Remember It for You Wholesale*

As the protagonist of Christopher Nolan's *Memento* reminds us: "Memory's not perfect. It's not even that good."

Memory does not capture every tiny detail for future inspection. Our brain is not a body camera collecting perfect video footage of our surroundings. It is a scrawled, handwritten note on a cheap paper napkin. And each time we recall an event, we are reading a photocopy of that napkin. Over time, only the boldest moments get recalled, while the details get fuzzy. Memories are

* One of the most memorable pieces of sci-fi tech in film is the neuralyzer. One look into the bright flash of light causes a short-term retrograde amnesia, useful for trying to keep the existence of extraterrestrials under wraps.

like an outdoor crime scene that decays over time, disturbed by the forces of nature. They are certainly not tattoos on the mind.

Dr. Elizabeth Loftus studies how eyewitness accounts are sensitive to outside influences, such as leading questions, and how these change the way a scene is remembered. In one experiment, participants watched excerpts from driver's education films showing traffic accidents. When given the prompt "How fast were the cars going when they *hit* each other?," on average, participants guessed 34 miles per hour. As if playing "Mad Libs: Law Enforcement Edition," the experimenters systematically changed the verb in their question. When presented with the more dynamic "smashed into," responses were closer to 41 miles per hour. However, when primed with the practically benign "contacted," witnesses thought the cars were going only 32 miles per hour. In another experiment, participants were shown an automobile accident video but subsequently prompted with two different questions, suggesting that the cars either "smashed into" or "hit" each other. They returned one week later to answer a follow-up question about the details of the crime scene: whether there was broken glass in the video. Most participants were correct in responding no. But nearly twice as many participants* incorrectly remembered this detail, with the "smashed into" condition more prone to falsely remembering fractured pieces of glass.

Another of Loftus's studies focused on a 14-year-old boy who was told four different stories from his own childhood by his mother and older brother. Loftus loves using games to demonstrate the malleability of memory, and this one is "Three Truths and a Lie." Over the course of the next few days, the boy was asked to journal about those events, providing as many details as he could remember. The false event given to him was the most barebones framework of a story: He got lost at a shopping mall, cried, was

* Sixteen out of 50 who read "smashed into" but only seven out of 50 who read "hit."

helped by an older man, and ultimately was reunited with his family. In writing about this entirely fabricated event, all sorts of specific details emerged, including information about his savior. Despite being completely fictitious, the helpful man wore a blue flannel shirt, was balding, and had glasses.

In the above examples, the harms are relatively small. Car accidents mean trouble no matter how fast the cars are going, and while gaslighting children is awful, a single instance with a substantial debrief afterward may not be that bad. The consequences of memory rewriting can be much bigger if these false memories produce heinous accusations with little to no concrete evidence. For example, psychotherapists have unintentionally implanted false memories during sessions. A patient might come in with symptoms characteristic of some common psychiatric diagnosis, maybe anxiety or depression. Of course, there are genuine cases of past childhood abuse or trauma, and when it happens, it increases the risk of all variety of mental health conditions. Occasionally however, in the process of speaking to patients about their childhood, perhaps under hypnotism, a therapist might accidentally implant memories of how the client was mistreated or neglected during childhood. Instances of this came to be called the "memory wars" of the 1980s and 1990s, which spread beyond psychology and psychiatry into the legal domain, resulting in malpractice lawsuits. Nadean Cool, for example, sued her psychiatrist for failed "therapy" after he diagnosed her with more than 120 multiple personalities, including angels, Satan, and a duck. Certain memories were implanted into her consciousness, including some where her father forced her to eat babies as part of cult rituals. In other sessions, her psychiatrist brought a fire extinguisher in case of spontaneous combustion during an exorcism.

According to the Innocence Project, more than 60 percent of their clients have been wrongfully convicted based on mistaken eyewitness identification. Some of them have served more than

35 years, and others are facing capital punishment, representing tremendous injustice and a failure of the legal system. The confidence of a witness is particularly sensitive to changing over time. For example, providing any kind of feedback during a police lineup may change the witness's demeanor while taking the stand during the trial. Looking at persons in the lineup, the witness might point to one of the people and say, "I think he's the one." If they receive any kind of confirmation, such as "That's who we had in mind," or subtle cues, such as a shift in body language, it rewrites the witness's memories, and their confidence skyrockets. Later, armed with that newly discovered assurance, whether genuine or inflated, they can be powerfully persuasive. To avoid this type of error, the lineup process must be modeled after a laboratory experiment. There is a hypothesis being tested (the witness can identify the culprit). There are control conditions (known innocent people, actors, or police officers who closely match the physical description). There should be no feedback provided to the witness during the lineup (a blinded test administrator, such as that used when the physician doesn't know which is the active drug and which is the placebo in a clinical trial). With this design in place, you can decrease the likelihood of false positives (the witness identifies the wrong person).

The list of cases in which Loftus was involved contains some of the most unpopular people of the past few decades, including O. J. Simpson, Ted Bundy, Jerry Sandusky, Bill Cosby, Harvey Weinstein, and Ghislaine Maxwell. She basically has the same role in all these cases, supporting the defense by suggesting that the witnesses who testified against them may have faulty memories. Naturally, this has made her plenty of enemies. She has been sued in the California Supreme Court for defamation and invasion of privacy. Academic speaking invitations have been revoked. She has faced the scorn of her law school colleagues at the University of California, Irvine, and has been shunned by other women for being a traitor to her gender. She has even received death threats. Nevertheless, she stands by her research as a reminder of the tenets of our legal system: "innocent until proven guilty" applies to everyone and even criminals deserve a fair trial.

Cobb: "What do you want from us?"
Saito: "Inception."

The fallibility of memory—*Total Recall*, *Inception*, or whatever you want to call it—is more real than anyone wants to admit. Loftus's work at the organism level has demonstrated that memory processes are sensitive to subtle influences, resulting in distortions of memory. These data have been backed by some remarkable findings at the cellular level in nonhumans, lending further support to the engram explanation of memory.

In 2013, just three years after the world was first treated to the visual and intellectual spectacle of Nolan's *Inception*, a landmark study in the world of memory implantation was published. The abstract opens with something well-documented since even before Loftus's 1974 study: "Memories can be unreliable." The rest of the article describes a series of experiments that sound like a barrage of science fiction tropes in an elaborate narrative dump. First, they replaced viral genes with a sequence that causes neurons to fire action potentials when struck by light. Then they injected that virus into the brains of genetically modified mice. They put these mice into chambers, each room uniquely equipped with their own characteristics, including lighting, room shape and size, composition of flooring material, and even different smells. A mouse would be placed in one chamber, which encodes a memory of that environment by activating a specific engram. The animal is brought into a new room, different from the original room, equipped with an electrified floor. A foot shock is delivered while activating the first engram, causing the animal to associate fear with the original engram. When put back into the first room, the animal freezes, the response when exposed to a fear associated stimulus.

The researchers incepted the memory of being afraid.

In the bizarro version of this experiment, researchers also implanted a positive memory by stimulating an engram previously associated with a good experience. Like a sudden dose of pleasant nostalgia at the flick of a light switch, reactivating that circuit reversed the set of depression-like symptoms that result from being stressed.

As with forgetting, we don't always get to choose which inputs get encoded. Sometimes, we unintentionally remember things we'd rather forget. Understanding the mechanisms behind this phenomenon also means understanding PTSD, a psychiatric condition affecting somewhere between 5 and 10 percent of the population. What makes PTSD such a puzzling memory disorder condition is that many people experience trauma, whether it be a near-death experience, severe physical injury, interpersonal violence, combat experience, or maybe an illness. But there is something different about the memory systems of those 5 to 10 percent of people who, in response to their trauma, experience frequent, recurring memories or flashbacks of the inciting event, complete with debilitating emotional and physiological responses. Exposure to associated cues may initiate an episode in the way a whiff of diesel fuel might instigate a response in combat veterans of the Vietnam War because it reminds them of the smell of napalm—morning, day, or night.

If the engram theory of memory holds true, PTSD might be a result of an overly active circuit in the presence of cues associated with a previous fear experience. Exposure therapy, one of the clinical strategies for helping patients with PTSD, might exert its effect through rewiring those fear engrams with something neutral or positive. Under the influence of a chemical that can enhance this cellular rewiring—psychoplastogens, such as ketamine, psilocybin, or cannabis—it might be easier to rewrite those painful memories.

CHAPTER SEVEN

"BORN AFTER THE SILENCE"

LANGUAGE AND OUR TWO BRAINS

GOOD SCIENCE IS PREDICATED ON REPLICATION. SOMETIMES, when studies fail to replicate, a closer inspection at the methods is warranted. So when a man gets the bumps on his head fondled as a means to predict his personality and gets two very different results, perhaps there isn't much "science" to this process.

On his first visit, he went incognito. The office was impressively decorated with busts, the borders of each little bump clearly delineated and labeled with different adjectives. He took a seat. The reader ran his hands over the man's skull uninterestedly, rattling off generic character traits: a spirit of daring, fearlessness, pluck, courage—all good things. Next was the reading of the other side of the head. Apparently, there was a giant hump here that corresponded to "caution," so large that all those other bumps were negated by the mountainous altitude of this bulge. Over and over, every good trait was negated by a bad trait. According to the skull reader, everyone has a bump of some size in one particular area, so you could imagine the surprise when his head had a relative indentation. Apparently, this was the first skull where this area was pushed in instead of bulging out. Consulting his many labeled marble busts to figure out what a dip meant, both were shocked to discover the character trait that should be lacking.

It was humor. The man, Samuel Langhorne Clemens—aka Mark Twain—who penned witticisms such as "God created war so Americans would learn geography" and the presciently self-aware "classic—a book which people praise and don't read," was found to be humorless.

Twain, now using his name and the impressive reputation that it carried, returned to the same office for a second reading. This time around, there was allegedly a lofty humor bump, the largest that the skull reader had ever felt.

Discrepant results such as these were common among practitioners of phrenology, one of the biggest pseudoscience fads of the 1800s. Before Twain* ran his experiment in the 1870s, the English-speaking world saw Franz Joseph Gall's massive manual of phrenology,† published posthumously in 1835. It was the culmination of Gall's work, a rationalization of the basics of phrenology using discussions of morality, comparisons between animal species, and philosophical principles. Even though it was written under the guise of a scientific work, it was informal observations of his childhood classmates that set him on the path toward phrenology. With more than a tinge of jealousy, Gall recalls how some of them were so much better at recitations, exerting far less energy than he. Of those who showed this talent, he noticed they had prominent eyes, which he attributed to an expansion of the brain areas that are responsible for this skill.‡

In Gall's view, contrary to the scientific consensus at the time, the outer surface of the brain was not just a single "brain." Instead,

---

* Twain never shied away from publicly making his distaste for phrenology known. As he wrote in *The Adventures of Tom Sawyer*, "A phrenologist . . . came—and went again and left the village duller and drearier than ever."

† The work had a comedically verbose title: "On the Functions of the Brain and of Each of Its Parts: With Observations on the Possibility of Determining the Instincts, Propensities, and Talents, Or the Moral and Intellectual Dispositions of Men and Animals, by the Configuration of the Brain and Head."

‡ While Gall respected excessively protuberant eye sockets, this facial feature is one that H. P. Lovecraft uses as a synonym for "weird" and "inhuman."

it was made up of several organs, each of which is responsible for a different mental function. If someone was particularly good at something, it was because their corresponding brain organ is bigger, and that increased volume will push on the skull, causing bumps that can be felt at the scalp by a trained expert. He categorized these organs as specializing in either intellectual faculties or feelings. The intellectual faculties are somewhat straightforward, including such features as perception of time, size, and weight. The feelings are more abstract, some of which included, using the original old-timey language, alimentiveness, amativeness, mirthfulness, and philoprogenitiveness.*

There were several flaws in the ways Gall and the phrenologists he inspired collected their data. For one, the ever-present selection bias tainted his sample population. Instead of studying participants who are representative of the whole population, he cherry-picked a narrow subset, including highly educated academics, people in the "insane hospitals," and the imprisoned, sometimes right before execution. As if working backward, he made observations about personality first and then tried to apply the measurements to the observation, violating the basic flow of hypothesis-driven science. In addition, these subjective observations were prone to confirmation bias. Basically, when you see something that agrees with your view of the world, such as a criminal with a large murder organ, you are quick to accept the results, thinking, "Oh yeah, this is normal and expected." But when you see something that contradicts your belief, you are critical of the finding or invent some alternative explanation. A child has protruding eyes but with an average memory? Maybe the memory test was poorly designed, or the child was just having a bad day. Let's just fiddle with the test a little bit more and test again in a month. Maybe the kid had an illness or brain

* Desire for food, amorous love, happiness, and a love for one's offspring, in that order.

injury or challenges in childhood that prevented the memory organ from working at its true capacity. The kind of mental gymnastics that phrenologists excelled at led to irreproducible science.

Parts of Gall's writing were revolutionary, and paradigm-shifting ideas made him enemies. According to phrenology, human behaviors such as religiosity are guided by some physical parts in the head, including brain centers for God and devotion. He suggested that when these areas meld with the adjacent murder organ, the person may be at risk of religious mania with the propensity for suicide. Most offensive to the Church was his claim that devotion comes from the interaction between concrete brain parts rather than the fuzzy, indefinability of a God-given free will. He also alienated many scientists, some who held steadfastly to the dogma that the cerebral cortex is a single, indivisible part of the brain and others who believed that we can learn only from external experiences, not things that come intrinsically from within.

Gall may have gone about the practice of science wrong, but he made claims that were right—or, at least, claims that align with our modern understanding of the brain. Importantly, he had voiced support for localization theory, the idea that different brain regions contribute to different functions. In support of this, he compiled a collection of brain injury case studies, showing that the resulting symptoms depend heavily on the location of the injury. Between the boy kicked in the head by a horse, resulting in spilled brain matter the volume of a hen's egg; the man who survived several years after being shot in the head with the bullet lodged against his pituitary gland; and the woman who was hit by a 30-pound rock, losing a fist-size amount of brain, there were no pronounced losses of their intellectual faculties.

While on the topic of case studies, a handful of clinical observations built the case in favor of brain function localization. What follows is the story of neurologist Paul Broca, who found a specific area of brain damage that was correlated with language disorders. In a convoluted, roundabout way, Broca provided a small bit of scientific support to what phrenology had tried to do.

> Obsidian jumped from the car, shouting. It was the first time Rye had heard his voice—deep and hoarse from disuse. He made the same sound over and over the way some speechless people did, "Da, da, da!" —Octavia Butler, *Speech Sounds*

Language contains a tremendous degree of nuance. It exists in subtle degrees; someone might be angry, bitter, cross, disgruntled, enraged, furious, or one of many other similar adjectives from the rest of the alphabet. We lie maliciously, carefully omit details, tell harmless mistruths, or deceive for the sake of developing the best psychology experiment. We unconsciously attach value judgments to words, intertwining history with sociopolitical biases, giving words the power to dehumanize and stigmatize. Language evolves rapidly, tethered to globalization, the rapidity of technological advancements, and the number of viral posts on social media.*

In 1861, at a hospital in the south of Paris, Broca encountered the patient who would later be described in nearly every psychology textbook. Although Broca formally addressed him as Monsieur Leborgne, the others in the hospital called him by his nickname, "Patient Tan." In the same way Olivia Butler's Obsidian could use only the syllable "Da,"† Patient Tan's vocabulary was limited to the sound "Tan"—of course, spoken with the silent "n" and the nasal twang of French. A careful examination of Tan's speech revealed

* Words added into the Merriam-Webster dictionary since 2021 include *hygge*, a mood of comfort and coziness taken from the Danish; *smishing*, a text message with the intent of gaining someone's personal information; and *yeet*, to throw with reckless abandon.

† Since I'm getting into the business of diagnosing fictional characters, I would also recommend George R. R. Martin's *Game of Thrones'* Hodor and Marvel Comics' Groot for neurological examinations.

that he used that syllable in place of other words, even using matching hand gestures and body language. He would speak with varying inflections appropriate to the emotion he was conveying: rising intonation when asking a question, a hushed "tan" when confiding secrets, and a rapid string of loud "TANTANTAN!"s when fearful or angry.

Like a detective examining every possibility, Broca began his investigation into the mystery of Tan's speech. He found no paralysis or injury to the complex set of throat, mouth, and tongue muscles required for language production. He assessed the extent of Tan's intellectual capacity and comprehension. How long have you been in the hospital? he would ask. Tan's nonverbal response: opening and closing his left hand four times, then extending the index finger. Five times four plus one, 21 years, an accurate count of his time at the hospital. Using this same approach, Tan could also tell the time, translating the numbers on a watch into left-hand gestures. His mental capacity wasn't diminished; he demonstrated comprehension, intelligence, creativity, and an ability to communicate, just not using spoken word. Broca concluded, incorrectly by today's definitions but accurate considering the evidence, technology, and terminology in use at the time, that Tan "lost their memory for words." He named this condition aphemia: a loss of speech.

Tan initially lost his speech at the age of 30, which brought him to the hospital. After years of care, Tan's condition began to decline. He developed a weakening of his right arm followed by the right leg. This weakness foreshadowed the eventual paralysis of the right half of his body, causing him to be bedridden for the last seven years of his life. In the end, he developed a severe infection of the right leg and died soon after. Broca ordered an autopsy immediately to maximize the freshness of the brain tissue. He elected not to cut up the brain using the crude mechanical methods of the day, instead opting to submerge the brain in a vat of ethanol. Without dissection, one abnormality was immediately evident. On the surface of the left hemisphere of the brain, there was a large, darkened hole about the size of a chicken egg. Gall may have arrived at the wrong conclusion about the localization of personality traits. But maybe, Broca thought, a behavior such as language might require the neurons found in this injured area of Tan's brain, which would account for why he spoke the way he did.

Then came along a second patient, an 84-year-old groundskeeper named Lazare Lelong. He too had a severe loss of language, but he was slightly better off than Patient Tan. There were five words at his disposal, including "yes," "no," "three," "always," and his own name. After Lelong died, Broca performed an autopsy, again removing the brain and submerging the organ in a second vat of ethanol. In Broca's words, "I will not deny my surprise bordering on stupefaction when I found that in my second patient the lesion was rigorously occupying the same site as the first."

One is a fluke. Two is a trend. Nineteen out of 20: now that's a pattern. Like a good scientist validating results through objective observations and multiple trials, Broca studied more patients with aphasia through postmortem autopsy, finding that often the most profound damage was in the same spot of the left cortex at the junction between the frontal, parietal, and temporal lobes, an area that would come to be called Broca's area. His conclusion: We "speak with our left."

Broca's awareness of the limitations of 1860s medicine paid off, and his prescience left a gift for modern anatomists better equipped to probe the brain without damaging it. Since then, new technology has taught us more about these language disorders, collectively called aphasias. While Broca could draw conclusions only about the areas of the cortex visibly damaged from the outside, magnetic resonance imaging (MRI) can digitally dissect the brain without harming the tissue. One hundred forty years later, MRI scans were taken of Patient Tan's and Lelong's ethanol-soaked brains, preserved reasonably well through the decades tucked away in a

Parisian medical museum. The results of the scans provided new context to Broca's observations. Both brains had severe damage of the left arcuate fasciculus, a long white matter tract connecting anterior and posterior brain circuits. Deeper frontal and parietal structures, which send and receive arcuate fasciculus signals, were lost as well, ravaged so thoroughly that they appear as gaping holes in the brain. The right hemisphere, by contrast, was structurally intact, an important comparison to meaningfully control for these brains so heavily affected by other disorders, such as seizures and late-stage dementia. Importantly, especially for phrenologists looking for evidence in favor of localization of function, the site of brain destruction was relatively specific. Many key brain structures, such as the cerebellum and hippocampus, as well as most other cortical areas, such as the occipital, superior frontal and parietal, and inferior temporal lobes, were structurally intact.

There have also been several case studies that extended Broca's observations about language. The location of the injury influences the nature of the aphasia. Patients with damage to the left inferior frontal gyrus (Broca's area) may present with symptoms such as those of Tan or Lelong. These patients speak in deliberate, stuttering phrases, often using only nouns and verbs while omitting tenses. Language comprehension is intact, and they respond in interviews with two or three words relevant to the questions. This specific diagnosis is expressive aphasia since their major difficulty is in their expression and use of language. Sometimes, the diagnosis is also called nonfluent aphasia, a reference to their disjointed language. Alternatively, it is also called Broca's aphasia in homage to his work.

An injury* that damages a different region, a little more toward the back of the brain in the superior temporal gyrus, leads to a dif-

* The most common cause of aphasia is stroke. About one-third of all strokes produce aphasia. Strokes are messy conditions, resulting in very diffuse patterns of brain injury. No two strokes are the same, from the blood vessels and regions of the brain affected to the severity of the damage.

ferent set of language-related deficits. These patients speak fluently, having no difficulty stringing words together or forming sentences. The issue is that the content of that speech is nearly meaningless, as if they are randomly pulling words out of the dictionary. A lot of the time, the grammatical structures may still be in place, using nouns and verbs where they are expected and so on. Sometimes, they exhibit paraphasia, or language output errors, as if the syllables or consonants are jumbled up somewhere right before they come out.* These patients also have a difficult time comprehending language. This language disorder is receptive aphasia, or fluent aphasia.

Importantly, aphasia is not just a deficit of speech, but of language more holistically. Ask people with aphasia to write a description of a scene, and the resulting text mirrors the symptoms of their aphasia diagnosis. For those with a nonfluent aphasia, they write mostly nouns and verbs, with squiggly letters, as if each stroke of the pen requires intense concentration. And in a fluent aphasia diagnosis, the writing is smooth, but the sentences lack meaningful content, as if words are strung together haphazardly. The same observation applies for manual languages such as American Sign Language, where a patient may use the correct arm movements and placement but with the wrong letter hand shape, akin to syllable switching seen in paraphasia. There may also be comprehension errors with Braille. In a case study of a blind 82-year-old German stroke survivor, his aphasia manifested as paraphasic errors in processing language associated with tactile cues, such as misreading *einladen* ("invite") as the nonword *geinmadel* or *kaisers* as *leider*.

> Are you getting any cross-chatter? . . . Between hemispheres. If there's damage to the left hemisphere, where the linguistic skills are normally located, then sometimes the right hemisphere will fill in to the best of its ability. —Philip K. Dick, *A Scanner Darkly*

* Instead of "language," paraphasic speech might sound more like "ganguage" or "lanluage."

There's a tremendous amount of variability between people even at the level of internal physical anatomy. Like Ian Fleming's fictional villain Dr. No, who survived a bullet to the left side of the chest where his heart should've been, some people are born with dextrocardia (the prefix "dextro-" means "right"). The more dramatic rearrangement of the organ blueprint is situs inversus, where all the viscera are mirror images of the "normal," textbook anatomy. People with these congenital birth conditions may not even know they have their organs oriented differently until they go for some kind of medical diagnostic testing* or, in some cases, during preliminary examination before an operation, much to the surprise of the surgeons. If the body sometimes builds the heart on the right or the appendix on the left, it's totally expected that sometimes the brain builds the language circuits on the right.

Because of this natural unpredictability between people and given that a good quality of life hinges so thoroughly on communication, it is critically important to avoid any injury that may damage the structures of language. In the context of neurosurgery, for example, brain tumors may best be treated by excising the offending tissue. Before the surgery proceeds, the surgeon might want to verify which hemisphere is mostly responsible for language to assess if aphasia may be a risk of surgery. To do so, they can take advantage of the way the circulatory system feeds blood into the two hemispheres fairly selectively using the left and right internal carotid arteries. The Wada test involves injecting a reversible anesthetic called sodium amytal† into one artery at first while the patient, fully awake during the procedure, is asked to perform a simple speaking task, such as counting. As the drug bathes over the

* Actress Catherine O'Hara (*Home Alone*, *Schitt's Creek*) had a routine electrocardiogram that produced some very strange results. Her physician prescribed a series of imaging scans and discovered that O'Hara is among the estimated 0.01 percent of people with situs inversus.

† The drug, also called amobarbital, has gained a reputation in entertainment as a "truth serum" because of its pharmacological action at inhibitory circuits. As one of the henchmen from *Ant-Man and the Wasp* says, "There's no such thing as truth serum. That's just nonsense from TV."

brain tissue, it temporarily shuts down that hemisphere. If it were the half that was responsible for language, the person would lose the ability to speak until the anesthetic is cleared away.

This left-versus-right assessment is good enough for surgeons. But research scientists, ever driven by a need to quantify phenomena, developed their own noninvasive method to assess just how strongly language is localized in one hemisphere or broadly distributed across the two. This simple and cheap test is the dichotic listening task. Participants wear a set of headphones through which two different but similar-sounding syllables, such as /ba/ and /ga/, are played at the same time into each ear. The participant is asked to report the sound they hear. Whichever syllable they say suggests which ear exerts dominance for language. Because of the way the ears are wired into the brain's auditory processing system, the right ear has a quicker neural route to the left hemisphere. At the end of the test, a laterality quotient is calculated, yielding a number between −100, total left ear (right brain) dominance, to +100, total right ear (left brain) dominance.

About 80 percent of people score in the positive range on this laterality quotient in dichotic listening tasks. Sex is the main moderator for differences in linguistic hemispheric dominance, with men typically scoring a higher laterality quotient, whereas women are more likely to score closer to 0. There is some evidence that handedness may be involved too, as left-hand-dominant people sometimes show lower laterality quotients. These observations track with clinical data, as language loss following left-hemisphere stroke is less severe in women and left-hand-dominant people.

Even at the time of *A Scanner Darkly*, people were interested in "cross-chatter": communication between the left and right hemispheres. Many neurons bundle their axons into thick white matter tracts called commissures, forming bridges of reciprocal communication between the two halves. The biggest and most prominent of these communication routes is the corpus callosum, made up of nearly 200 million neurons and spanning about 10 inches long in the adult brain.

There are people out there without this interhemispheric communication. Some forms of epilepsy produce atonic seizures, which lead to sudden falls and head injuries, and radical surgical approaches are sometimes recommended if the symptoms are not treatable by lifestyle changes or medication. One such operation is to sever the corpus callosum. The theory behind how this surgery helps is the same way you would stop that ear-piercing screech when you hold a live microphone next to a loudspeaker. Epileptic brain activity is like a sound that gets amplified repeatedly, distorting the pitch and increasing the volume. The seizure signal in one hemisphere travels across the corpus callosum, triggering another wave of excess activity before returning to the original hemisphere and repeating the process. Severing this communication pathway stops this cycle of positive feedback from happening, thereby eliminating (or at least dampening) the severity of the epilepsy.

The study of these "split-brain" patients has provided clues about the role of interhemispheric communication. Granted that all goes well in the operating room, they typically fare better in the long run because the surgery lowers the risk of epilepsy attacks and restores or improves quality of life. The side effects of the surgery are surprisingly minimal, considering the hundreds of millions of communicative cables that have been severed. There may be problems with coordination of large muscle groups as well as the smaller muscles involved in speech. Sometimes, there are memory or language deficits. The effects on the brain are often temporary, especially in pediatric cases, where they commonly resolve within a few months.*

* Probably, their relatively plastic brains are better able to cope with the surgery due to compensatory changes in other white matter tracts, of which we have at least 21.

The most striking side effect of damage to the corpus callosum is alien hand syndrome. This is when a limb, usually the left hand, performs autonomous but purposeful actions beyond the person's will or intent, producing the sensation that the hand is not their own. For example, the hand may gently stroke the face and hair while watching TV. Elsewhere in the medical literature, alien hand syndrome is also called anarchic hand syndrome or Dr. Strangelove syndrome.* Alien hand syndrome leads to intermanual conflict, where the two hands will bicker over the best way to do the same task. A steering wheel may be turned in two opposite directions, or the two hands will wrestle for control over a dexterity task. Like Thing from *The Addams Family*, some alien hands have their own personalities and preferences. The patient might go to the dresser to get a light cardigan, but their alien hand may quickly grab a wool sweater instead. Some people personify their alien hand, often using negative valence words ranging in implication from the mischievous "cheeky" to the more sinister† "the devil's." One woman believed her alien hand was a newborn named Joseph who mischievously bites at her during faux nursing.

Patients overcome the alien hand by sitting on it.

> CLOSE SHOT—ANIMATED HAND—(7 SEC.) The hand throws the trap from itself and raises its middle finger to ASH, flipping him the bird. —*Evil Dead 2*

Following a callosotomy, there may be long-lasting language deficits but only under highly specific experimental circumstances. In these experiments, split-brain patients were asked to focus their

* In reference to Stanley Kubrick's 1964 dark comedy about the Cold War. *Doctor Strangelove or: How I Learned to Stop Worrying and Love the Bomb* shows a character with a right hand that occasionally does the "Heil Hitler" salute on its own will, much to the dismay of the reformed ex-Nazi.

† Choice of words here is important: "sinister" comes from the Latin word meaning "left," which has been thought of as evil.

vision on a small plus sign in the center of a screen. Then two different images would blink quickly on the right and the left sides of the screen, maybe a house on the right and a face on the left. If you ask them, "What did you see?," they will say, "House." If you put a marker in their left hand and ask them to draw what they saw, they would draw a face. If you ask them, "Why did you draw that?," they'll look down at the face and respond, bewildered, "I dunno."

It's almost as if one body is controlled by two different pilots.

Here is a brief aside to describe the neuroanatomy at play in this response. The nervous system exhibits contralateral organization; that is, stuff in the right sensory field (house) is processed by the left half of the brain. For most people, the signals are sent across white matter tracts seamlessly between the left and right hemispheres, so both hemispheres quickly learn about both the house and the face. But for the split-brain patient, this communication doesn't happen. In accordance with Broca's observation, the left brain excels at language, so the person can verbalize whatever it knows, responding, "House." However, the left sensory field (face) is detected by the right hemisphere, which is typically less equipped for language, so the person doesn't mention the face. But the right hemisphere has dominion over the left half of the body—again with the contralateral organization—so when the right half of the brain sees the face, that's all it knows, so it tells the left hand to draw a face. The person then becomes confused when asked about their drawing because the verbal left brain now sees the right brain's drawing and becomes conflicted: I said "house," so why did I draw a face?

This left-brain dominance for language is also observed for other nonvisual sensory systems. The sense of smell, as a notable exception to much of mammalian neuroanatomy, sends signals into the brain using ipsilateral connections; that is, smells entering the left nostril are processed by the left brain. In agreement with the language-in-the-left-brain schema, patients can better name odorants when the right nostril is plugged and an odor is wafted beneath the left nostril. Somatosensation, the sense of touch, like vision, has a contralateral organization, and objects placed in the right hand are processed by the left brain, so stimuli can be identified by words. But without interhemispheric communication, a split-brain patient who cannot see their left hand holding a ball would not be able to say they are holding a ball, nor will they be able to pick up a different ball with the right hand.

Again, one body, two pilots.

For split-brain patients, their separated hemispheres learn to communicate with each other in ingenious ways characteristic of the creativity of the brain(s). This is called cross-cuing. One patient, JW, was shown a number to his right brain. His right hand, controlled by the left brain, which didn't see the number, would point arbitrarily to a number on a number line. Then the left hand would point up or down. JW's two brains were playing the "Hi-Lo" game using JW's body as the intermediary. In working with a different split-brain patient, the testing procedure had to be modified mid-experiment because her cross-cuing was interfering with the results. At the beginning, she pointed quickly to targets, often guessing. After a few trials, she changed her strategy. There was a noticeable pause, followed by a slight head movement and then a point. She was using her chin to point to the correct target based on the knowledge that one hemisphere had. In a separate "Name the Smell" experiment, a person couldn't name a disgusting aroma wafting under only the right nostril. But certain scents elicit an unconscious, bilateral facial reaction, such as the universal "yuck" face, characterized by brow furrowing, nose wrinkling, and corners of the lips pointing down. Then the vocal left hemisphere might recognize the feeling of these muscles moving and offer guesses fitting for the emotional response. Is it hydrogen sulfide, the eggy-sulfurous smell of a sewer? Or maybe it's butyric acid, the smell of vomit? Patients consistently guessed better when trusting their facial expression.

Split-brain patients are often described as a monolith, especially in introductory-level neuroscience and psychology textbooks. However, many of these results are often broad simplifications of the complete set of observations. In the final paragraph of the first split-brain report published in 1968, the author* acknowledges that there were occasional inconsistencies with results and even some outright exceptions. Of the 11 patients studied, the extent of the surgery differed, which may account for the discrepancies. Some had a complete commissurotomy, the more radical surgery where the corpus callosum along with other major interhemispheric white matter tracts were also transected. On the other hand, some patients had only a partial callosotomy, usually leaving the back one-third of the corpus callosum intact. There were also confounding variables complicating the picture, such as the fact that all these patients had epilepsy and were likely on a regimen of high-dose anti-seizure medications, both of which change brain activity broadly in complex ways.

Behaviorally, much evidence points to the left hemisphere as being stronger at processing language. It is much more difficult to study language at the cellular or molecular level. Some clues about the role of genetics in language came about from the KE family. Many members were diagnosed with apraxia, a disorder characterized by expressive and receptive difficulties as well as late language acquisition. Some members did not speak their first words until they were seven years old. The appearance of apraxia in roughly one half of the 30 family members across four generations suggested an autosomal dominant pattern of inheritance. Further molecular analysis located something interesting on chromosome 7 in a

* Roger Sperry went on to share a Nobel Prize for some of this work.

region called FOXP2, thus earning that sequence the nickname the "language gene."

As with most things in genetics, FOXP2 does more than just language, and many genes other than FOXP2 are used for language. But it was the first gene strongly implicated in the process of language acquisition. The gene provides the instructions for making the protein FoxP2, a molecular gatekeeper positioned to control the fate of many other proteins. As a transcriptional regulator, it binds directly to specific sequences of DNA, thereby preventing that section of DNA from being accessed by the machinery that turns DNA into mRNA and then protein. FoxP2 contributes to neurite outgrowth, the developmental process essential for making sure that neurons form connections with the appropriate communication partners. Accordingly, one hypothesis suggests that affected members of the KE family had a faulty FoxP2 protein, leading to improper neuronal development during the critical window for language acquisition.

A major challenge with studying language is the lack of a good experimental nonhuman model, one in which scientists can carefully control the many environmental variables that would be impossible to control for in people. Language as we know it is an exclusively human trait, which puts major limitations on the types of experimental studies that can be performed without violating ethical standards in research. Of course, nonhumans communicate, from dancing bees and ants faithfully tracking a pheromone trail to baby mice that emit high-pitched squeals when looking for attention to the musky odor of female deer when signaling reproductive readiness. But in these examples, animal-to-animal communication is more rudimentary, conveying concepts far simpler than what we can with our words. Furthermore, these behaviors are mostly innate or instinctual rather than learned and refined through practice, as language is.

One model organism that displays some similarity to human language learning is the zebra finch (*Taeniopygia guttata*). Small enough to fit in the palm of your hand, these birds have brightly colored beaks and spots and stripes across their tiny fluffy bodies. Language scientists are most interested in their pleasant chirping melodies. Despite having literal bird brains with millions of years of evolutionary pressures favoring smaller and lighter brains, they exhibit a language-learning capacity with surprising analogies to human speech. Their songs use a vocabulary of specific patterns of changing pitch, inflection, rhythmic pauses, and volume. They string together notes and syllables into motifs just like humans put phonemes together to make sentences. Within a month after hatching, they rehearse a primitive, not-yet-formed song made up of softer sounds, akin to newborns babbling. Birdsong plays a major role in mate attraction, and they learn this song best when receiving feedback, such as the fluff-up dance that mothers do when the juveniles successfully mimic the vocal patterns of their fathers,* just like how human babies and toddlers learn better with frequent rehearsal and positive reinforcement. As the birds mature, they improvise and adapt their song until they develop a unique vocal signature. They can recognize and distinguish other finches based on their individualized songs, much in the same way humans can identify and differentiate James Earl Jones from Kristin Chenoweth based on vocal pitch and timbre.

A region of the zebra finch brain essential for their singing is Area X. Artificially manipulating levels of FoxP2 changes the shapes of neurons in Area X and the birds' ability to learn and accurately reproduce a song, implicating this region as the bird analogue of a localized structure for language. These researchers are embroiled in debate about the mammalian analogue of Area X since it shares cellular, circuit-level, and neurochemical similar-

* Get out of here, Freud.

ities with different parts of the human brain. But nearly all agree that Area X is part of the basal ganglia, a network of brain areas in humans involved in habit formation and motor control, both of which are key elements of language.

> The illness, if it was an illness, had cut even the living off from one another. As it swept over the country, people hardly had time to lay blame on the Soviets (though they were falling silent along with the rest of the world), on a new virus, a new pollutant, radiation, divine retribution. . . . The illness was stroke-swift in the way it cut people down and strokelike in some of its effects. But it was highly specific. Language was always lost or severely impaired. It was never regained. —Octavia Butler, *Speech Sounds*

Could an illness cause localized brain injuries, bringing on a dramatic loss of language as described in Butler's postapocalyptic tale? Possibly.

Start with the herpes simplex virus. Scarily, epidemiological data estimate that most people worldwide harbor some herpes viral load; many live their entire lives not even knowing. The crafty virus infects neurons by traveling across synapses,* seeking refuge in the brain, where they can hide from the roving forces of the immune system. For many, viral infection is asymptomatic. When symptoms do appear, the most well known are the painful and unsightly cold sores, which lie on the mildest end of the spectrum. A small percentage of cases progress to the clinically dangerous herpes simplex encephalitis. As with most brain-squeezing conditions, such as meningitis or hydrocephalus, patients experience general symptoms, such as fever, vomiting, confusion, and, in severe cases,

* Scientists, an equally crafty bunch, have harnessed this neuroinvasive property of similar viruses to visually dissect the connections in the nervous system. Researchers have replaced the genetic material of the normal herpes virus with a sequence with codes for fluorescent protein. The virus then delivers that code into neurons, causing the cells to make a bright signal, which lights up the neuron, showing exactly where that neuron projects to.

a loss of consciousness or seizures. Survivors of the encephalitis sometimes have such severe damage to their brains that it can be seen using brain-imaging scans. The pattern of injury is typically across the temporal and inferior frontal lobes, right where Broca's area resides, explaining aphasia after recovery.

Another infection-related brain lesion was described by Viennese psychiatrist Constantin von Economo in the early 1900s. World War I took at least 20 million lives, but many of them were not a consequence of direct armed conflict. A major scourge of public health was the viral influenza pandemic of 1918, historically called the Spanish flu.* A particularly virulent strain of the flu, estimates put the number of total infected close to one-third of the global population with up to of 100 million deaths as a direct result of the illness. Granted, epidemiological record keeping was low on the list of priorities, coming in after the restructuring of geopolitical lines, maintaining economic stability, and rebuilding society.

Following shortly after the major spike in influenza cases were reports of a mysterious brain-swelling disease that left more than half a million fatalities between Europe and the United States. Named encephalitis lethargica, patients suffered from many of the classic symptoms of brain-swelling conditions. What distinguished this condition from other forms of encephalitis was the profound lethargy that the patients experienced. Many would sleep excessively; some were even comatose. More than a million people were afflicted by the disease, which came to be known as "sleeping sickness."† Instead of the trademark intractable somnolence that gave the disease its name, one in 10 of these patients had the completely

* The practice of naming diseases after places is a lazy way to portray other people as less hygienic (read worse) than people from one's own country. In most cases, the nomenclature is just factually wrong. The "Spanish" flu probably came from Kansas.

† Confusingly, also the name given to a completely different illness. The other sleeping sickness is caused by a parasitic infection transmitted by the tsetse fly. In late stages, this other sleeping sickness causes psychiatric symptoms including hallucinations and delirium, sensorimotor problems, and an inversion of the day–night sleep cycle: daytime drowsiness combined with nocturnal insomnia.

opposite sleep profile: this minority of patients were completely unable to sleep.

Von Economo had divided his patients based on the two sleep patterns. And when they died, as about one-third of them did within the first few years (or sometimes days) of their diagnosis, he went to work in the pathology lab. Removing the brains from the patients' skulls, he found them soft and bloated with water, necessitating an extra step of tissue preservation before cutting them into slices. Dissection revealed that there was a correlation between the nature of their sleep pattern and the area of damage in their diencephalon, the region of the brain directly underneath the cortex. If the patient presented with the prototypical sleepiness, there was a notable lesion in the posterior region. On the other hand, for that 10 percent of people with the never-sleeping sickness, the damage was in the anterior area. He concluded that encephalitis lethargica produced specific injuries of the diencephalon that dictated the changes to their sleep behaviors. These examples demonstrate that there is precedence that a viral infection could lead to highly focused injuries of the brain, and they would smother the world in a blanket of speechless silence if they had a particular penchant for targeting language circuits.

Butler also supposes that the blame might fall on a pollutant if not a disease. The most famous example of a chemical causing a highly localized area of brain damage, at least among my little circle of neuroscientists, is the case study of the "Frozen Addicts." In the 1980s, there was a sudden appearance of a novel neurodegenerative disease throughout Northern California. Neurologist Dr. William Langston was surprised to find six people presenting to emergency rooms with a similar set of symptoms akin to late-stage Parkinson's disease. They moved slowly if at all. When they did move, their limbs twitched jerkily. There was a near-constant muscular rigidity, so their limbs were stiff and resistant to movement. Some were even "locked in," unresponsive for hours at a time, just looking out of their emotionless faces with a blank stare. Their condition worsened exceedingly fast, and people were losing their movement over the course of weeks instead of decades. This was all very confusing for Langston. These patients were young and therefore didn't fit the profile of a typical patient with Parkinson's disease, who often start developing symptoms in their sixties.

Through some clever detective work, Langston solved the mystery. The rapid speed of onset suggested toxin exposure rather than a body breaking down by some degeneration. Before the California outbreak, there was the 1976 case of Barry Kidston. The chemistry graduate student performed some experiments in his garage to synthesize the designer drug MPPP, an analogue of pethidine, one of the most prescribed and recreationally misused opioid medications at the time. Like a sloppy Jesse Pinkman from *Breaking Bad*, a tiny mistake in the process resulted in an impure product containing trace amounts of a related chemical called MPTP. This accidental by-product is not the inebriant that people sought: it was a potent neurotoxin.*

To be precise for the sake of any pedantic biochemists out there, MPTP is inert and harmless. It's neurotoxic only when it is processed by one of our enzymes called monoamine oxidase (MAO). MAO helps break down chemicals such as dopamine, norepinephrine, and serotonin, turning them from neurotransmitters into inactive by-products. However, when MAO acts on MPTP instead, the resulting product is MPP+, which is the actual neurotoxin. Its chemical structure contains a free radical, a rogue unpaired electron, which has a nasty habit of plucking electrons out of other molecules, like an atomic home-wrecker. MPP+ interferes with mitochondria, thereby disrupting the energy supply that these

* This exact neurotoxin was alluded to by William Gibson in his cyberpunk classic *Neuromancer*. "Condition's like Parkinson's disease, sort of," says Molly.

power-hungry neurons need for survival, leading to rapid dopamine neuron–selective cell death.

Certainly, the leakage of a neurotoxin into the drug supply was tragic for the 100 or so patients exposed, many of whom developed these Parkinsonian symptoms. On the upside, there was some good that came of all this pain and suffering. Like language, it is very difficult to study Parkinson's disease experimentally. Laboratory animals never develop Parkinson's disease on their own. You can take a population of nonhuman primates, age them up until their fur is speckled with gray, and even extend their life spans with the best possible monkey health care, and they still show no paw tremors or postural and gait changes characteristic of human Parkinson's disease. Developing an experimental model for Parkinson's disease would allow for the testing of new medications and therapeutic strategies in a nonhuman setting rather than relying on risky human experimentation. It turns out that injecting MPTP into experimental primates also produces rapid-onset degeneration of their dopamine cells, manifesting as tremors and other early signs of Parkinson's disease. This primate model for Parkinson's disease has been critical for the development of new medications and surgical approaches to treat people with the condition.

The Michael J. Fox Foundation has a portion of their website dedicated to Langston, MPTP, and the case of the Frozen Addicts.

CHAPTER EIGHT

"BELIEVE ONLY HALF OF WHAT YOU SEE"

ILLUSIONS AND DELUSIONS

OUR BRAINS CAN BE AMAZINGLY POWERFUL COMPUTATIONAL tools, capable of flexibly adapting to difficult or changing situations. But sometimes, it glitches out while attempting to process our surroundings, resulting in an imperfect impression of the world. Maybe it's a random burn mark that looks strangely like a cat hanging from a noose. Maybe it's the sound of a thumping heartbeat getting louder and louder while police scrutinize the house. Or maybe it's the phantom vibratory buzz of your cell phone in your pocket.* From paleolithic times, when protohumans were scratching images onto both sides of a disc and spinning them rapidly to produce the illusion of movement, to the viral image of the blue and black dress (or was it white and yellow?), we are fascinated by tricks of the mind.

> I gathered that what he chiefly remembers about it is a horrible, an intensely horrible, face of crumpled linen. —M. R. James, *Oh, Whistle, and I'll Come to You, My Lad*

* The chapter title is an adage often attributed to Edgar Allan Poe, the author who described all of these examples in his writing—except for the obvious anachronism.

Pareidolia is the natural propensity to interpret vague stimuli as something else more easily contextualized. A fluffy white cloud is a mass of condensed water vapor, not a camel, a weasel, or a whale. Led Zeppelin's "Stairway to Heaven" played backward is gibberish, not a secret tribute "to my sweet Satan," as Christian TV producers in the early 1980s would have you believe. Small elevation changes on Cydonia, Mars, are rocky plateaus captured by low-resolution photography, not a sculpture of a face or evidence of ancient Martian civilizations covered up by NASA scientists.

If it weren't for the ominous warning engraved on the ancient whistle in M. R. James's story, the perception of a face among the abstract shapes of wrinkled bedsheets is not supernatural or even the slightest bit unusual. Two shapes just above a third centrally located shape often evoke the impression of two eyes and a mouth.* Not only do we mistakenly believe that inanimate objects are alive and staring back at us, but we also ascribe emotions to the "being." Power outlets, with their small, rounded mouths, look surprised. Car headlights above an oversized vertical grille have a mischievous Cheshire Cat grin. A semicolon followed by an end parenthesis mark gives the impression that the book is winking seductively at you.

;)

So why are we quick to imagine faces among inanimate objects? One proposition suggests that detecting faces is highly adaptive, so it is therefore beneficial for the visual system to err on the side of false positives. Faces were so important for protohumans since early hominids were much hairier than us and therefore may have looked similar from the neck down. Sure, traits such as physique, posture, and gait were different between individuals, but

* A woman from Florida made an easy $28,000 in 2004, when she auctioned a piece of grilled cheese with a "divine" image of the Virgin Mary on it. She stopped eating the second she noticed the Blessed Mother's "face" in the darkened grill marks of the toasted bread, so the famed piece of leftover lunch is missing a bite-size chunk. As further evidence of the divinity of the grilled cheese in question, the seller claims that it has never grown mold despite being more than a decade old on sale.

facial structures offer far more discriminative power. The best way to differentiate the safety of kin from the danger of strangers was to quickly examine tiny features on the face, such as the distance between the eyes, the width of the nose, the curve of the lips, and the angle of the jawline. These neural circuits responsible for facial recognition probably adopted as many shortcuts as possible, reducing the cognitive energy required to distinguish faces from non-faces—hence the spooky bedsheet.

Dr. Justine Sergent dedicated much of her research career to studying a neurological impairment called prosopagnosia, where a person is unable to recognize others by looking at their face. They have no deficits with the anatomical structures; they know you have two eyes, a nose, and a mouth, for example. Some of them may be able to identify eye color. But their difficulty becomes apparent when they see you a second time, maybe later in the day. They may not recognize you. In severe cases, they don't recognize the faces of their loved ones or family members or even their own face in a mirror. And while prosopagnosia can certainly lead to all sorts of social embarrassment, many people avoid these situations by using context clues. Non-facial features, such as your fashion sense, posture, hair, subtle mannerisms, and the quality of your voice, can be used to identify you. Some people can also recognize certain characteristics on your face, such as scars, freckles, acne, or subtle asymmetries. In a way, prosopagnosia is the conceptual opposite of facial pareidolia: not enough faces or too many faces.

Sergent was the first to anatomically identify the pattern-recognition machinery involved in the processing of faces. To do so, she used an experimental technique called positron emission tomography (PET) scanning. In this live brain-imaging technique, the researcher uses radioactively labeled compounds to figure out which areas of the brain get increased levels of blood flow during different conditions. Sergent put participants in the PET scan and showed them a series of images, including faces, objects, and gratings. She saw a specialized clump of brain cells found in the ventromedial aspect of the right temporal lobe at the base of the brain that are more engaged during face-related tasks, such as recognition of previously seen faces or classifying them based on perceived gender. One such area came to be called the fusiform face area (FFA) in reference to the stimuli that most robustly change their activity. This research was extended* by other cognitive neuroscientists who showed that just thinking about faces activates the FFA and that injuries to the FFA lead to prosopagnosia. Also, the FFA is activated more by faces than by houses—unless the house has two open windows above a central door, evoking the pareidolic sensation that the house is alive. This trick of the visual system was employed by the designers of the cover art for Mike Flanagan's *The Haunting of Hill House*, which depicts the titular mansion and its twin lighted windows superimposed over a woman's face.

Because of facial pareidolia research, the work of Renaissance painter Guiseppe Arcimboldo saw a revival in the world of brain imaging. In his time, a new art style called Mannerism was taking root. The focus of this movement was less about producing exact replicas of the subjects and more about highlighting the most glamorous features of the subject. While his work with Holy Roman emperors and other members of high court garnered him status, it was his extracurricular endeavors that he is most remembered for in the neuroscience laboratory. In these pieces, Arcimboldo used a collection of organic objects to re-create faces. *Vertumnus* is a face constructed out of pears, cucumbers, and an ear of corn for, well, the ears. *Spring* is made of daisies and roses, while *Autumn* is made of pumpkins and grapes.

---

* Sergent's groundbreaking research career ended just a few years after her characterization of the FFA. The academic harassment probably started long before an anonymous letter was sent to the press accusing her of putting her research participants at unnecessary risk. The harassment escalated until Sergent tragically died by suicide on April 11, 1994. No evidence of ethical infractions was ever discovered.

Figure 8.1. *The Jurist*, a portrait of a sleazy politician, was constructed from the surprised mouth of a rotting fish, the pallid flesh of turkey, and the backside of a greasy, misshapen bird. GIUSEPPE ARCIMBOLDO, PUBLIC DOMAIN, VIA WIKIMEDIA COMMONS.

Arcimboldo's paintings have been used as diagnostic tools to study different aspects of facial recognition. Some patients with prosopagnosia do not see the faces in the paintings, reporting individual plants or vegetables instead of seeing the big picture. Infants spend more time looking at important stimuli compared to less salient stimuli, and Arcimboldo's paintings hold the gaze of seven-month-old infants, indicating they find faces more interesting than assorted vegetables.

> A strange multiplicity of sensations seized me, and I saw, felt, heard, and smelt at the same time; and it was, indeed, a long time before I learned to distinguish between the operations of my various senses. —Mary Shelley, *Frankenstein*

In the first few moments of existence, Frankenstein's creation describes synesthesia. Derived from the Greek *syn-*, meaning "union," and *-aesthesis*, meaning "sensation," synesthetes experience a simultaneous blending of senses. In their worlds, activation of one sense triggers the activation of an unrelated sensory system. For example, musician Pharrell Williams, on the opening track of N.E.R.D.'s third studio album *Seeing Sounds*, describes his encounter with the surreal experience that inspired the album name. Neurologist Richard Cytowic writes about a synesthete at a dinner party in his book *The Man Who Tasted Shapes*, embarrassed that his roast chicken dish lacked not salt or pepper but "points"—it tasted too spherical, too round—and was therefore unfit to serve. Hearing shapes, smelling colors, or tasting sounds are everyday experiences for synesthetes.

Instead of a one-to-one connection between the sense organs and the relevant processing areas of the brain, the wiring diagrams of the sense organs in a synesthete might have been assembled by an absent-minded switchboard operator. For whatever reason, one line gets connected with other pathways, resulting in an unexpected three-way call. According to Cytowic, there are more than a hundred different unions of senses. Fuzzy, abstract concepts (emotions or locations in space) and human-designed constructs (days of the week) may be subject to this perceptual mixing. According to one of his patients, Wednesday is the color indigo blue. Some synesthetes feel that a sequence of numbers in ascending order don't align on the straight edge of a ruler but instead along a roller-coaster–like track, complete with rolling hills, precipitous drops, and loop-de-loops. For yet others, speaking with introverts may evoke a refreshing sensation, as if a cool bucket of water were splashed on their face on a hot summer day.

According to many self-reports, synesthetes are either ambivalent toward or enjoy the sensations. And thankfully so for a couple of reasons. First, synesthesia is inescapable. These multisensory experiences are involuntary, meaning there is nothing that can be done to suppress these sensations. Second, their experience is consistent over the course of their lifetime. Once a synesthete starts seeing "A" as red, they will perceive that Letter as Scarlet for the rest of their lives.

The most common synesthetic experience is grapheme-color synesthesia, where letters or numbers evoke the perception of colors. The intensity of this sensation lies on a spectrum, with lesser forms evoking only the vaguest thought of a color, while full-blown synesthetes see a whole rainbow of colors, as if every letter was printed in a different hue and luminance. While there are no hard-and-fast rules about synesthetic pairings, they are guided at least in part by environmental experiences and psycholinguistic rules. In fact, the above example of "A" being red is far more common than if the colors and letters were matched purely arbitrarily. One scientist suggests that this association may have been created or at least reinforced by the red "A" that appears in the set of magnetic alphabets that were common in preschools. Another common trend is for the first letter of the color name to bring about that color: "B" is often blue, and "Y" is often yellow. Vowels are often less bold than consonants, sometimes evoking white, yellow, or transparent. When asked about words as a whole, the color of the first consonant or vowel is often also the color of the word: the color of "cat" is likely also the color of the stand-alone letter "c" or "a." In multisyllabic words, the emphasized syllable determines the word's color, and pronouncing the word while stressing the syllabary in unusual ways can change the perceived color.

The first clinical description was penned by Georg Tobias Ludwig Sachs in his 1812 medical dissertation. Originally in Latin, he describes the senses of a brother and a sister, both born with albinism. For them, letters, numbers, musical notes, ideas, and days of the week produced the impression of colors. He noted that context was more significant than the shapes of the symbols on the page since the number 1000 and the year 1000 produced different colors. Although Sachs's dissertation was carefully written to maintain scientific distance from the patients in question, it was autobiographical: He was the brother.

The tools of modern neuroscience labs still haven't solved the mystery of why synesthesia happens. For one, there's not a tremendous amount of time or resources dedicated to it. These experiences are certainly unique, but they aren't life threatening. Synesthesia is not considered pathological and is not predictive of any adverse health consequences (although there is an elevated comorbidity with anxiety disorder and OCD). Still, curious clinicians and scientists have unsurfaced a few clues about the origins of synesthesia.

For many synesthetes, their experiences begin in early childhood at just a few years old, roughly around the time when they are starting to grasp abstract concepts. For example, quantities up to 20 (fingers + toes) or even in the hundreds range might be feasibly countable, but imagining what a thousand or a million of something looks like forces the imagination to stretch beyond what is easily manipulable, somehow leading those sensory circuits to connect in atypical ways.

The neural changes required for synesthesia may start even farther back than this, before we even take a breath of real air. For most people, the nervous system develops connections according to a generic, default plan that guides the way the nervous system is wired. Instead of the complex branching shapes of adult neurons, immature neurons are more round with only a handful of short projections. To make the right connections in all the right places, neurons play a cellular version of the "hot or cold" guessing game. Neurons probe their surroundings by temporarily extending parts of their cells, tentatively groping around in the dark. Sometimes, they come across signals that repel the neurite. Other times, different proteins beckon the neuronal projection, stabilizing the cellular skeleton, as if to say "you're getting warmer!" This process is important for guiding the axons as they weave their way throughout the young nervous system. If these signaling molecules differ in subtle ways, it could result in a brain wired in unexpected ways, such as an auditory cortex directly connected to the color-processing region of the occipital lobe or a gustatory cortex linked with the dorsal stream of the visual cortex. We have more brain connections over the first few years of life that gradually fade away. But in synesthetes, these connections might persist, leading to a permanent "miswiring."

For example, consider auditory-visual synesthesia, the chromesthesia that fuels Pharrell Williams. Researchers hypothesized that there might be stronger connections between areas V4/V8, two of the key associative structures of the visual cortex that process color, and the medial and superior temporal gyrus, which are auditory associative areas that process sound. To test this theory, they conducted a brain scanning study using diffusion tract imaging (DTI), a variant of the MRI. In DTI, the scanner is set up to study the movement of water molecules. Consider a molecule in the middle of a glass of tap water. That molecule can diffuse in any of the three dimensions freely with no impediments to movement. But a water molecule in the brain is influenced by nearby cellular structures, more likely to move in the same direction as axons rather than against them. This tendency to "go with the grain" can be quantified in a measure called fractional anisotropy. Areas with high fractional anisotropy therefore contain more white matter. In agreement with their hypothesis, the inferior fronto-occipital fasciculus, a major white matter tract that connects the visual and auditory structures with the frontal lobe, has a greater fractional anisotropy in the right hemispheres of synesthetes compared to control participants.

Some of the genes that code for axonal guidance proteins, such as ROBO3 and SLIT2, explain at least partially why synesthesia sometimes appears in families. Nearly 40 percent of synesthetes have at least one first-degree relative with synesthesia, although their particular perceptual blend is likely unique even between identical twins. Both Vladimir Nabokov, author of *Lolita*, and his son are synesthetes. For Nabokov, letters have colors and textures. "M" is pink flannel, "D" is a creamy yellow, "P" is an unripe apple, and "KZSPYGV" makes a perfect rainbow. Singer-songwriter Billie Eilish, her dad, and her producer brother Finneas are all synesthetes. According to Eilish, her 2019 breakthrough song "Bad Guy" is reddish-yellow, evokes the feeling of the number 7, and smells like cookies.

And if it seems like many of the above-mentioned synesthetes are famous artists, being able to perceive the world in an unusual way might give them a slight edge in the ridiculously competitive field of creative careers.* Animator Michel Gagné depicted his own synesthetic migraines while working on *Ratatouille* when the

* Although synesthetes may leverage their unique sensory experiences as a device for creative inspiration, it doesn't always work to their favor. Some sound-feeling synesthetes struggle with composing since certain combinations of notes elicit profound physical discomfort, forcing them to work without certain tools in their repertoire.

culinarily inclined rodent bites into a strawberry and sees a dazzling light show of dancing flames, exploding patterns of radiating colors, and flowing neon lines. Jazz pioneer Duke Ellington could see timbre differences and could identify which saxophonist was playing based on the colors he heard. Grammy Award–winning composer Leonard Bernstein (*West Side Story*) uses his chromesthesia to compose with both sound and sight, literally painting pictures with notes. Hans Zimmer, who composed the expansive soundscapes to some of modern science fiction's greatest films, including *Inception*, *Interstellar*, and *Dune*, attributes his inspiration to a combination of his lack of formal classical training and his ability to see sounds. Olivia Rodrigo chose purple as the backdrop for the cover art of her debut album, *Sour*, because many of the songs, including "Driver's License," are purple to her. The names of creative synesthetes make up quite an impressive Liszt.*

To alleviate your mounting sense of FOMO over these unique gifts, there are ways to experience synesthesia even if your neural circuitry isn't naturally wired for that function. But, as a warning, some of these gateways to synesthesia have huge costs and the trade-off may not be worth it:

- Lysergic acid diethylamide (LSD, or acid) and other mind-expanding psychedelic drugs: The most common self-reported perceptual blend during intoxication is mild visual sensations evoked by sounds. Since most of these classic psychedelics exert their influence through activation of serotonin signaling pathways, differences in these circuits might be a cause of congenital synesthesia. However, possession of these drugs is still a federal crime in most of the United States.

* Franz Liszt, Hungarian composer from the Romantic era of the mid-1800s, was a color-sound synesthete. When conducting, he would sometimes ask his orchestra to play certain passages more loudly, with more emphasis, or while using more blue. Only the last instruction would be met with confusion.

- Hemorrhagic stroke of the thalamus: After recovery, a patient acquired strong sensory-emotional synesthesia, reporting that words printed in blue evoke a strong feeling of disgust. He also reported that loud, high-pitched brass instrumentation, such as the iconic octave leaps at the beginning of the James Bond theme song, evoked an "orgasmic" sense of euphoria.* The self-report was backed by objective fMRI data that showed a pattern of widespread brain changes consistent with high positive emotional arousal rather than the relatively localized pattern seen in the ordinary pleasurable enjoyment of music. The subnuclei of the thalamus receive inputs from many of the sensory systems, so the damage from a stroke might have caused a compensatory rewiring of those once-segregated pathways.
- Temporal lobe brain tumor: A patient with a long-lasting headache, hemilateral weakness, and new-onset auditory-visual synesthesia where sounds evoke comet-like streaks of light went in for a medical evaluation. Standard neuro-ophthalmic examination discovered intact, healthy visual pathways. Other tests discovered a brain tumor. Lifesaving surgery to remove the glioma stopped the synesthesia.
- Amputation of an arm or leg: These people are prone to developing a mirror-touch synesthesia, where seeing someone else being touched produces the sensation as if they are the ones being touched. However, they are more likely to feel that touch sensation in their missing phantom limb rather than the original body part observed.

If you're not willing to break a federal law, experience life-threatening brain damage, or sacrifice a body part just for a taste of

* Not every scientist enjoys a pop culture reference while reading academic publications, but I do. This case study was described in a paper titled "From the Thalamus with Love."

synesthesia, let me offer up a consolation prize. Most people, likely you as well, have a hint of a shape-sound synesthesia. Consider the following shapes. One of them is named Bouba, while the other is named Kiki. Which is which?

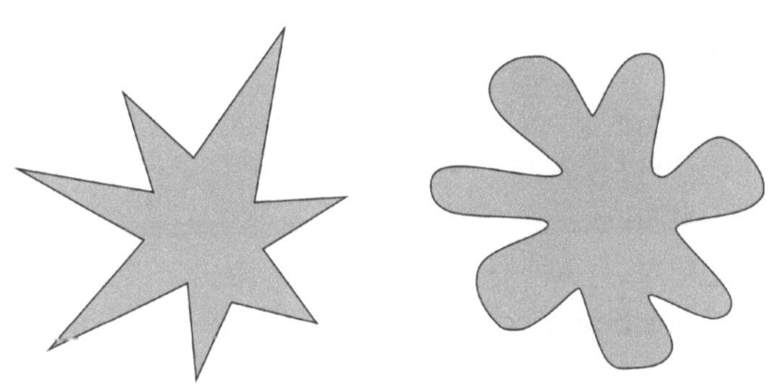

Figure 8.2. Monochrome version, June 1, 2007.
BENDŽVECTORIZED WITH INKSCAPE—QEF (TALK) 21:21, JUNE 23, 2008 (UTC), CC BY-SA 3.0 <HTTP://CREATIVECOMMONS.ORG/LICENSES/BY-SA/3.0/>, VIA WIKIMEDIA COMMONS

If you felt that the angular spiky shape on the left is Kiki while the rounded feller on the right is named Bouba, you would be among the 70 percent of people who feel that way. It is sometimes called the Maluma/Takete effect, the original names given to the round/spiky shapes in Wolfgang Kohler's 1929 experiment. More recent follow-ups to this phenomenon observed a similar effect across many cultures, even those using writing systems with non-Roman scripts, such as Korean, Japanese, Thai, and Greek. This effect is even seen among the Himba people of northern Namibia, whose language has no written component and who have practically no exposure to English or other Western or Eastern scripts.

Synesthesia is fully in the realm of the healthy. The next few sections are examples of cases of illusions and delusions that appear in illness.

> However, this bottle was NOT marked "poison," so Alice ventured to taste it, and finding it very nice, (it had, in fact, a sort of mixed flavour of cherrytart, custard, pine-apple, roast turkey, toffee, and hot buttered toast,) she very soon finished it off. "What a curious feeling!" said Alice; "I must be shutting up like a telescope." And so it was indeed: she was now only ten inches high, and her face brightened up at the thought that she was now the right size for going through the little door into that lovely garden.
> —Lewis Carroll, *Alice's Adventures in Wonderland*

On one hand, we have Scott Carey from Richard Matheson's *The Shrinking Man*, Stan Lee's superhero *Ant-Man*, the children from the live-action Disney comedy *Honey I Shrunk the Kids*, Matt Damon's character from *Downsizing*, and, of course, the well-mannered and curious Alice Liddell.

On the other hand, we have their real-life counterparts, who feel as though they have shrunk to a six-inch-tall figurine suddenly dwarfed by the giant wastebasket in their room or that their room has suddenly expanded into a football field–length hallway, like they just sipped from Lewis Caroll's cherry and turkey–flavored drink. At other times, they feel the opposite sensation. As if they had just eaten the cake conspicuously labeled "EAT ME," they feel like their point of view had elongated until they are far above their hands, that they could effortlessly reach out and touch the opposite wall, or that their head expanded to twice its size like some comically disproportionate Thanksgiving parade balloon.

These are case descriptions of Alice in Wonderland syndrome. It is more often reported among children than adults and is associated with many other conditions, most commonly as part of the auras that precede a migraine attack* but also in epilepsy, while

* According to Lewis Carroll's journals, he likely experienced "odd optical affections" right before the onset of a "bilious headache," suggesting that his own distortions of reality were a source of inspiration for Alice's adventures.

battling infectious diseases, or following exposure to medications and psychedelics.

Researchers haven't arrived at any definitive mechanism that accounts for these sensations. Naturally, the earliest theories about these experiences fell prey to Freudian psychoanalysis, the great gravity well that influenced so much of brain science of the early 1900s. According to this interpretation, the shrinking sensation is a manifestation of regressive behavior, a defense mechanism whereby a person's behaviors are those of an earlier stage of their development. So maybe they imagine themselves shrinking so that they can be cared for and loved unconditionally like an infant, thereby escaping the challenges that are inherent to growing up.* Evidence-based approaches have failed to discover obvious structural brain abnormalities in these patients, but studies have seen decreases in blood flow and unusual EEG activity in the occipital and parietal lobes. Both observations align with our understanding of brain organization, considering that these lobes contain circuits involved in visual perception and representation of the self, respectively. Each shred of evidence does explain a little bit, but the story is getting, as Alice would say, curiouser and curiouser.

An altered perception of other people can be equally disturbing. One of the biggest mouthfuls in all of the neuroscientific literature,† prosopometamorphopsia is a visual illusory condition where a person sees faces that slowly change their shape, as if they were looking at the infamous trickster of Norse mythology Loki. Some of the terrifying visions seen by people with this extraordinarily rare condition include the following:

* Although a true Freudian might suggest that shrinking has to do with worries over impotence or genital size.

† In first place is dysdiadochokinesia, an impairment of motor control where a person has difficulty performing alternating actions rapidly, such as flipping the hand back and forth on top of the other hand.

- Facial features that bulged and shrank disproportionately, as if every face were a conglomerate of individual beings haphazardly glued together.
- The face of a doctor whose kindly visage transformed into that of a terrible witch, glasses resting atop a long, crooked nose. Her teeth grew into a pair of fangs.
- The left half of faces stretched vertically with a giant eye, while the right half of the face stayed mostly normal—except for a blackened hole where the right eye should have been.
- "Melting zombie faces" that merged into a single, shrouded entity that hovered over the person as they lay in bed.
- Disjointed faces, made up of jagged, fractured fragments, with facial structures appearing where they do not belong.*
- Blank eye sockets on a misshapen face that scowled menacingly. And if one face gives this effect, what about another turn of the screw: The patient saw two of these faces as a mirrored pair, back-to-back, akin to the Roman god Janus.
- The skin of others' faces slowly turned progressively blacker and scalier, the nose grew longer and pointier, and eyes changed into brightly colored red, green, or yellow orbs. For the entirety of this woman's life, she saw ordinary people transform into dragon-headed people.

Cases of prosopometamorphopsia are often secondary to other illness, injury, or epilepsy that disrupt the normal activity of circuits passing through temporo-occipital areas. Damage to the corpus callosum, the major left-brain/right-brain interhemispheric communicative pathway, also may produce these face-morphing hallu-

* Based on this clinical description, it might be reasonable to think that Pablo Picasso's cubism period may have been inspired by prosopometamorphoptic hallucinations, although this theory is in the realm of speculation.

cinations. This suggests that we process facial features as a whole rather than as a piecemeal collection of eyes, a nose, and a mouth and that when some element of this gestalt understanding of faces is damaged, the individual pieces get distorted.

> At last, it reached this extremity, that while I was reading to the congregation, [the monkey] would spring upon the open book and squat there, so that I was unable to see the page. —Sheridan Le Fanu, *Green Tea**

Alice in Wonderland syndrome and even prosopometamorphopsia are twisted distortions of what is already, blurring the line between reality and the unknown. True hallucinations, seeing things that are not there, erase this line completely. Considering how dependent humans are on the visual world to orient ourselves with our surroundings, apparitions of living beings we see while awake are a truly disturbing twist of reality. They take on many shapes: a black-furred monkey perched on a Bible, a kindly child psychologist, or a gremlin on the wing of your airplane.†

Also known as visual release hallucinations, Charles Bonnet syndrome is when a person with severely worsening vision perceives complex fictive images. In 1760, Bonnet described a case study of his own 90-year-old grandfather, who experienced progressive vision loss following cataract surgery. The man started seeing all manner of things, ranging from people and animals to buildings and carriages. Recent studies suggest that this type of hallucination affects some 20 percent of people with severe vision loss—which is likely an underestimation, considering a hesitancy to sound "crazy"—making it a common occurrence among those

* Here, a devout clergyman is haunted by a hallucination of a black monkey who follows him around.
† *The Twilight Zone* episode "Nightmare at 20,000 Feet" stars William Shatner, who feels threatened by this being who disappears whenever someone else looks out the window.

with visual impairment. In addition to those visions described by Bonnet's grandfather, other reported hallucinations include floating shapes, vivid color patterns, vast landscapes teeming with plants and flowers, and childhood cartoon characters.

Sometimes, patients experience Lilliputian hallucinations, a reference to the society of six-inch-tall men on the fictional island of Lilliput described in Jonathan Swift's *Gulliver's Travels*. For most of the cases, these tiny people twirl and dance with their oversized velvet hats and skirts, perform acrobatics, and march in neat formations. Sometimes, they ride bicycles and play tiny musical instruments. Other times, they are accompanied by a menagerie of animals, ranging from cats, tigers, and hippopotamuses, all appropriately scaled down in size. They don't always produce auditory hallucinations, but when they are heard, everything about the voices and sounds are appropriate for their size, higher in pitch, and softer in volume with a Lilliputian timbre. These illusions don't always cause distress: for one patient with depression, these tiny fictive companions were "his only joy in life."

Even with the evidence provided by modern methodology, the neural substrates underlying these hallucinations have not been fully disentangled. One clue comes from people's subjective experiences under the influence of DMT, the chemical isolated from *Banisteriopsis caapi*, or the "vine of the ancestors" for the peoples indigenous to the Amazon rainforest. When mixed with other plants and brewed into a tea called ayahuasca, DMT produces a short-lasting and intensely strong psychic effect, ranging from somatic changes, such as a buzzing or tingling sensation, to profound emotional and cognitive changes. It also produces visual hallucinations. According to self-reported anecdotes, people see a variety of entities, ranging from jesters, deities across the panoply of gods and goddesses, beings made of light or fractals, tiny machine elves, and tentacled jellyfish-like beings. Some describe "a tree of life and knowledge," "a ballroom with crystal chandeliers," and the more ominous "huge fly eye bouncing in front of my face" and a "little round creature with one big eye and one small eye, on nearly invisible feet." The interactions were mostly good, but a few people reported being torn apart or being consumed by their hallucinations.

DMT interacts with the brain in several complex ways, although one of its key sites of action are at the serotonin 2A (5-HT_{2A}) receptors. These receptors are found in many different circuits across the brain. One key circuit regulates activity of the thalamus: the processing structure that incorporates signals originating from several different sensory systems, including the somatosensory, auditory, and visual organs. According to this thalamic filtering model, DMT activates the 5-HT_{2A} receptors, modifying the signaling pathway and resulting in aberrant and unexpected activation of a visual circuit. The thalamus also receives strong inputs from a region of the brain stem called the raphe nucleus, the main origin of serotonergic neurons, providing further evidence to support this theory.

Seeing is believing. But believing on its own, even without seeing anything? Just having the sensation as if someone is there can be just as reality-bending as seeing or hearing apparitions. These spooky feelings are called presence hallucinations. These felt presences vary in just as many ways as other hallucinatory experiences. The patient may feel as though they are visited by someone unidentifiable; other times, they know the presence personally ("my late ex-boyfriend standing in the middle of the room"). They might feel as though the presence is interacting directly with them in combined somatosensory tactile hallucinations, maybe feeling as though they were touched gently on the back of the head. Sometimes, the presence is "malevolent . . . moving towards me and I can't move or look at it or even scream"; other times, the feeling is neutral, "not necessarily for my benefit," while at other times, the presence has a purpose: "I had succumbed to hypothermia in a race and had basically decided to sleep in some marshalls and give up. Something told me to open my eyes, and ahead I saw a light. I sprung back into life and found my way back to the finish of the race. If I hadn't, I would be dead, no question about it."

A particularly poignant case study in these hallucinations is that of Dr. Daniel Drubach, a neurologist with Lewy body dementia at the age of 60. One of his many symptoms is the feeling of an illusory companion whom Drubach had named the "Presence." He never saw the Presence, but he got the impression that it was an older man, "unkempt but clean," with long gray hair who walked with his hands behind his back. In dementia, there are often waves of symptoms, ebbing and flowing periodically. And when the dementia scaled back, the Presence would also disappear, prompting Drubach to search for his companion. At times, he has even conceived of the Presence as an angel sent by God to protect him: Once, the Presence had pushed him back onto the sidewalk after he nearly walked into a busy street.

Like most illusory experiences, presence hallucination does not have a clear mechanistic explanation. A hint comes from a study that opens with a startling sentence: "Stimulation of a site on the brain's left hemisphere prompts the creepy feeling that somebody is close by."

In this 22-year-old woman, a grid of stimulation electrodes was placed close to the cortex, allowing the researchers to activate small sections of the brain. The temporoparietal cortex in particular prompted that feeling of presence sensation. First, when a small stimulation was delivered, she reported feeling a person, in her own words, a shadow. "He is behind me, almost at my body." A stronger stimulation was given, and she now felt that shadow man embracing her, arms wrapped around her waist. While laying down, she felt someone touching her left thigh. None of these sensations were described as being the feeling of a loved one. Instead, they were deeply unsettling.

Then she was asked to perform a simple language task, one where she flipped through a deck of image cards while naming the items depicted on the cards. Like the black monkey from Le Fanu's story, the illusory man interfered with her performance of the task. "He wants to take the card," she says.

> They topped the second hill, and then the path sloped through a headhigh swatch of bushes and tangled underbrush. It narrowed and then, just ahead, Louis saw Ellie and Jud go under an arch made of old weather stained boards. Written on these in faded black paint, only just legible, were the words PET SEMATARY.
> —Stephen King, *Pet Sematary*

In the same way *The Monkey's Paw** centers its vaguely hinted-at horrors regarding the lengths a grieving parent will go in the face of tragedy, *Pet Sematary* plays into the fear of loss. The other big underlying fear that powers tales of supposed reanimation is that of being buried alive, a situation so expertly described in many of Edgar Allan Poe's works, notably his *The Premature Burial* (1844) and more recently shown in Quentin Tarantino's *Kill Bill: Vol. 2* (2004). In Poe's time and beyond, an occasional misdiagnosis would lead a living patient to be interred when they simply had an undetectable pulse and depressed respiration. Several days later, when they spontaneously reawaken, they may find themselves in the original sensory deprivation chamber, the one six feet under, smothered in silence and total darkness.

And even in the face of all this, there are people who are absolutely convinced of their own death. Some even go to great lengths by voluntarily laying themselves down in the narrow confines of a coffin. This is a complicated belief, one that we will

* W. W. Jacobs's 1902 short story about the magical trinket that grants three wishes but at a terrible price.

come back to later. Sometimes, this and similar beliefs are a result of a disorder called dementia praecox, a catchall term for any premature psychiatric illness, coined in 1852 shortly after Poe's death. In more modern terminology, some of those cases would be diagnosed as schizophrenia.

One out of 100 Americans will be diagnosed with schizophrenia at some point in their lives. Despite the high occurrence of the disorder and the millions of dollars of research funds distributed annually toward understanding schizophrenia, there are still a lot of questions about the disease. For one, we don't exactly know what causes it. There are a few hints that schizophrenia probably results from some genetic factors since the risk of developing schizophrenia is slightly higher if someone else in the family has the disorder* and identical twins are more likely to share the disease than fraternal twins or siblings. Environmental factors also increase disease risk, as both maternal hardship and early heavy drug use lead to a greater chance of developing schizophrenia later in life. We have no idea what causes some people to develop the disorder and why others are resilient. There are a few hints about how brain chemistry differences may be involved in schizophrenia, and these scraps of knowledge have led to some medications that can decrease the severity of schizophrenia. But science is far from understanding the scope of the big picture.

What is known are the symptoms that plague patients with schizophrenia. They sometimes have psychomotor movement difficulties, such as catatonia, the holding of the body in entirely unnatural positions for prolonged periods of time, or stereotypies, purposeless repetitive movements. They frequently have auditory hallucinations, the so-called voices in their heads, talking to them and telling them things. People with schizophrenia also often

---

* The most famous person with schizophrenia might be John Nash, the Nobel Prize–winning creator of the field of game theory and the protagonist of *A Beautiful Mind*. Both he and his son have schizophrenia.

have difficulty organizing their thoughts, resulting in unusual, difficult-to-understand speech. Sometimes, this is a result of occasionally interrupting the flow of communication with a response to their auditory hallucinations or thinking that their own voice is externalized, as if speaking to themselves.

Some of the strangest psychiatric symptoms are delusions, untrue beliefs that the patient believes are absolute, unwavering truth. These delusions are so persistent that no amount of logical reasoning or scientific evidence can dissuade a patient from their belief. Sometimes, the nature of these beliefs hinge on highly implausible content. Delusions manifest in a few unique ways:*

- Delusions of persecution: Persecutory delusions are the most common form of delusion present in schizophrenia. These delusions can cause someone to take on several false beliefs that "the world is out to get them." Others believe that they are being followed constantly as paranoia invades their daily routine. Shades of persecutory delusions can be seen in horror fiction whenever some unwitting protagonist goes mad from being doggedly followed by something awful, whether perceived or real.
- Delusions of grandeur: People with these delusions have an overexaggerated belief of their own power, influence, or importance. Perhaps they believe they have God-like omnipotence and are therefore burdened with the obligation to save humanity from a great evil. Some believe they have supernatural gifts, maybe the ability to talk to animals, foresee the future, or read minds. Some grandiose delusions are a bit more grounded in scope, where a person might

* While delusions occur with schizophrenia, they are not exclusive to the condition. They may stand alone or be related to other complex psychiatric illnesses, such as OCD or manic episodes in bipolar disorders. The remaining examples of delusions in this chapter are not necessarily comorbid with schizophrenia.

believe they are royalty and must be addressed as such or are part of a critical intelligence-gathering mission while On Her Majesty's Secret Service. These beliefs sometimes lead to social ostracism or, worse, immediate physical harm. One patient, Sophie, while reflecting on her delusional beliefs after treatment, recognized the dangers of the actions she took while insisting she was Jesus. "In some cases I wouldn't think through where I tried [walking on water]. So maybe it will incidentally be shallow . . . but also in deeper places."

- Delusions of control: Another delusional experience is the belief that the person is being controlled or manipulated. Patients report feeling as though some malevolent puppet master is pulling their strings. Maybe it's agents of the experimental psychic branch of the government, an alien entity from beyond the stars, or a revenant of a wronged person from their past. They may report that every thought or idea has even been implanted by some external force. The protagonist of H. P. Lovecraft's *Shadow Out of Time* describes this delusion of control:

  > *Had my present body been the vehicle of a frightful alien consciousness from palaeogean gulfs of time? Had I, as the captive mind of those shambling horrors, indeed known that accursed city of stone in its primordial heyday, and wriggled down those familiar corridors in the loathsome shape of my captor? Were those tormenting dreams of more than twenty years the offspring of stark, monstrous memories?*

The experiences of people with schizophrenia, both the voices and the delusions, are often inspired by sociocultural and religious beliefs. Christians talk of God, Jesus, Satan, and exorcisms; Muslim patients hear voices of jinnis or *azan*, the call to public prayer; and adherents of Buddhism blame karma for their mental state. A transcultural study examining auditory hallucinations found that the common representation of the voices as being angry and encouraging acts of destruction or self-harm is not universally the case. In Christian patients from Accra, Ghana, the guiding voice is that of God, benevolent and kind, telling them to ignore the voices of the Tempter. One patient described a conflict between an evil serpent sent by a wicked sea witch and a guardian angel, the latter of whom eventually banished the snake back to hell. After that triumph of good over evil, the voice of angels was all she heard. Patients from Chennai, India, a society where kinship is highly valued, often hear the voices of ancestors or those of family members. Some offer calming advice, even reminding the person to take a bath, brush their teeth, and refrain from smoking and drinking. The voices of patients from Los Angeles suggest a different course of action, encouraging doing things such as "cut off their head" and "drink their blood."

Entertainment, pop culture, and the news also make up some of the influence over delusional content. A fascinating subset of the persecutory delusion is the belief that everyone is a paid actor or that the patient lives in an artificial environment where every action is being televised for the world to see. Perhaps every piece of dialogue has been scripted or they need to starve themselves to look good for the camera. This delusion has been given the unofficial diagnosis of Truman Show syndrome, named after the 1998 film starring Jim Carrey. These diagnoses started appearing in the early 2000s, coinciding with the global popularity of television franchises such as *Survivor* and *Big Brother*. Much more recently, occasional case reports have explored the delusional content that has been driven by SARS-CoV-2 and the related societal changes during a pandemic. A young Greek man, Mr. K, believed he was the personification of COVID-19, arriving as a plague to threaten humanity. Another man developed persecutory delusions that he was being trailed because he had the cure for COVID and desperately needed to reach a doctor to pass on the secret.

Interpersonal family dynamics, as complex and fickle as they can be, sometimes also guide delusional content. Firstborns are more likely to experience grandiose delusions, perhaps burdened with a need to serve as the protector over their younger siblings. On the other hand, youngest children are more likely to experience persecutory delusions, possibly a twisted extension of being looked after by others during those key, formative years of childhood.

Social hallucinations, such as voices, hallucinations of people, or the feeling of a companion, may be prompted by loneliness. According to one theory, when a person lacks the interpersonal relationships that are an ingrained component of our species, the brain may create fictive companions, even at the cost of twisting reality. This social deafferentation theory is in line with predictions made about Charles Bonnet syndrome: if the loss of sight predicts visual hallucinations, then the loss of interpersonal interactions may predict hallucinations with social content.

Think back to that person who insisted on burying themselves. Could a delusion lead someone to so fully believe they were dead? Consider the following self-reports and case studies and decide for yourself:

- A woman avoided showers, citing the fear that her fragile, rotting flesh would melt under the stream of water.
- A middle-aged woman discontinued her antidepressant therapy and soon experienced overwhelming nausea, which she reported was in response to the scent of her own flesh, rotting away to the bone.
- An elderly man was admitted to the hospital due to complications resulting from his severe anxiety over a concern that it had taken several years for him to be buried.
- A young man woke up hurriedly one morning, late for a funeral. He scrambled into his most formal black suit, with

matching black button-down and black slacks. He left his house and ran several miles toward the church where his own funeral was to be held.

The written history of this psychiatric disorder can be traced back to a 1788 manuscript by Charles Bonnet, the same man who described visual hallucinations in blind patients. Here, Bonnet described an otherwise healthy 70-year-old woman, suddenly hit by a stroke one evening as she was preparing dinner. After the incident, she became partially paralyzed and could no longer speak. Her family did everything to help her recover. When her voice finally returned, she revealed that she was irate at her friends and family. How dare they allow her to go on for this long without a proper burial. She insisted to her family repeatedly that she had died on that day when the stroke happened. Naturally, her family feared her unusual behavior, and, of course, they did not accede to her frequent requests to be buried alive. She violently threatened her maids until they gave in. They eventually prepared a coffin for the living, breathing woman. Before being placed in the coffin, she double-checked the pins in her hair, the straightness of the seams in her funerary shroud, and the whiteness of the linen sheets along the inside of the coffin. Even the living dead have a sense of fashion! And, of course, after some time, she would emerge from the coffin as alive as when she was first "buried." The process of being laid down in a coffin did little to shake her delusion. On waking, she would eventually get the feeling of being dead again, ask to be buried, and then surprise herself with a miracle when she came back to life. And this cycle repeated multiple times. Alive, dead—coffin. Alive, dead—coffin.

Almost 100 years later, the condition was given its official name by French neurologist Dr. Jules Cotard. He observed a similar set of symptoms in one of his patients, a woman who was anonymized with the name Mademoiselle X. She insisted that she had no blood, no nerves, and no internal organs at all. She believed her body was made up of only skin and bones controlled by a soul that was destined for eternal damnation. As a result of her delusion, she believed there was no value in eating. This delusion was so powerful that it eventually overwrote hunger, one of the most basic survival instincts that comes preprogrammed in the human body since birth. Soon after diagnosis, she died from self-starvation. Dr. Cotard named Mademoiselle X's condition *délire des negations*, a delirium of negation. She, Bonnet's stroke survivor, and the other patients described above who reported that their skin was rotting away or that they had died or were entirely incorporeal all are some variation of the delirium of negation.

Another name for this rare delusion, the one used when a neurologist really wants to make headlines in the popular press because of an eye-catching diagnosis, is Cotard's delusion, or "walking corpse syndrome."

CHAPTER NINE

NO LONGER DR. JEKYLL

The Changing Nature of Personality

A METAL ROD, THREE AND A HALF FEET (1.1 METERS) LONG, rocketed through the head of Phineas Gage, shattering the base of his skull and destroying a huge chunk of the man's brain. His survival was miraculous on its own. But it was the dramatic change in his personality that secured him a place in every introductory psychology textbook. As described by his friends and acquaintances, he was "no longer Gage."

I'm skipping to the punch line here. Let's go back to the first page of the story.

Along the eastern edge of Vermont, a construction crew had been tasked with demolishing a path through the Green Mountains to build a new railway. A likable young foreman named Phineas Gage oversaw that crew. Capable, efficient, and driven, he was known among his employers as an energetic, hard worker. With the acumen of a wise businessman, Gage always completed his contracts and always conducted himself professionally. His friends spoke highly of his well-balanced and active mind, and he displayed a temperance beyond his years. Terraforming in the mid-1800s was a brutish physical ordeal, done mostly with powerful, volatile explosives. The grass and dirt were stripped away until only rock remained. Small holes were drilled into the earth and then packed tightly with explosive powder, which focuses the massive destructive power of the blast directly into the surrounding rock. To best compress the explosive powder, someone would use a tamping iron to pack a layer of inert sand and a fuse.

On September 13, 1848, when Gage forcefully rammed his tamping iron directly into the explosive powder without the buffer of protective sand, he became a victim of a horrific workplace accident. The friction from the metal rod meeting rock set off a spark, one hot enough to ignite the explosive powder. The sudden expansion of matter propelled the 13-pound (six-kilogram) rod straight through his head, effortlessly punching a 1.5-inch (four-centimeter) hole clean through his skull. The tamping iron was later recovered some 65 feet (20 meters) away, covered in blood and brain matter.

All of the 1,000 people in the nearby town of Cavendish probably heard the explosion. Despite the ragged state of his body after the blast, Gage was not as shocked as you might expect. He stood up unassisted after the explosion and walked toward a nearby cart for transportation into the city. He remarked that he "does not care to see his friends, as he shall be at work in a few days." Gage demonstrated his classic sense of humor and trademark wit when he first met the physician, remarking of his condition, "Here is business enough for you!" The prognosis was grim. The hole was bleeding profusely, his left eye was completely obliterated, and Gage was literally dripping brain material onto the doctor's floor. Over the next few days, he was closely observed by Dr. John Martyn Harlow. Without proper antiseptic procedures, the inevitable infection progressed quickly. Gage remained in a semicomatose state for a few weeks while Harlow treated his injury with crystals of silver nitrate, rhubarb, and castor oil.

Less than a month after his accident, Gage's health started turning around. He became more active. He started talking more frequently and eating more heartily and was no longer in pain. When he finally gained the strength to get out of bed, he was interviewed by Harlow about the events of that fateful September 13. Not a single fact was lost from his memory, and his capacity for language was perfectly intact. He could recount details such as the time of day, what happened, and how he was wheeled from the construction site to the hospital by cart. Considering the severe damage to Gage's head, his recovery time was expedient. By the end of November that year, he had returned to his home, and by the start of the new year, the hole at the top of his head had fully healed.

Although his physical recovery was quick, his friends and family noticed that everything was not completely back to normal, mostly because he wasn't behaving like his usual self. When the tamping iron shot through his skull, it took a part of his personality. Instead of the friendly man who was once adored and respected by his coworkers, this new Gage was a profanity-spewing, irreverent, and "pertinaciously obstinate" man. The responsible part of him had been destroyed, and he was unable to create plans and stick to them, incapable of maturely handling advice given to him. He acted impatiently with no restraint, utterly capricious and almost animalistic. He had also lost his business acumen. He was once offered $1,000 in exchange for a few pebbles he took from the worksite. He declined.

Records of his life post-injury are hazy, largely lost to time. This new Gage, with his poor sense of responsibility, was unfit for the dangerous work on the railroad construction crew. After he was let go from his job, some say that he became a vagrant, that he would appear at Methodist services claiming his survival was due to divine intervention, or that he worked as a farmhand in California or a stagecoach driver in Chile or appeared on behalf of the more famous Phineas of the time—P. T. Barnum—as an exhibit in his New York museum. He died 11 years after his accident due to complications related to seizures.

Today, Gage's skull and the fateful tamping iron can be seen at the Warren Anatomical Museum at Harvard Medical School. Although the soft tissue of Gage's brain is long gone, some might argue that modern imaging techniques provide even more information about his condition than what the original examinations offered. The trajectory of the rod was re-created using skull X-rays and computer modeling. Some pathways were ruled out on the basis that major blood vessels would have been destroyed or that the ventricles would have been compromised, leading to massive infection; in neither case would the injury be survivable. Based on the most likely course of the projectile, the brain damage was limited to the white and gray matter of both left and right hemispheres of the frontal lobes. Broca's area and the supplemental motor areas, corresponding approximately to language production and motor control abilities, were untouched, in alignment with the observation that those functions were intact post-injury. The most severe damage, however, was to the prefrontal areas. Injuries here are consistent with Gage's deficits in emotional processing, impulse inhibition, and rational decision making.

Phineas Gage reminds us that underneath every Dr. Jekyll is a Mr. Hyde. It's just a clump of cells in the frontal lobe that holds him back.

I'm not a murderer. I've never even met the man I'm supposed to kill. —*Minority Report*

Personality changes can appear under less extreme circumstances than Phineas Gage's brain injury. There is often a recommendation for brain imaging in the case of new-onset behavioral changes, especially when drastic personality shifts are noted. In support of localization hypotheses, the behavioral and psychiatric symptoms that appear following the formation of a brain tumor sometimes depend on where the tumor is located. For example, tumors of the hypothalamus, the neurosecretory brain structure involved in regulation of our hunger drive, may produce anorexia. A pituitary tumor sometimes leads to new psychosis. And in agreement with Phineas Gage's lesion, frontal lobe tumors may make people act more recklessly. For most people, their personality defines them. The thought of some uncontrollable patch of cells in the brain subverting their entire being is on the same level as the horror of mind control.

In 1966, there was the case of Charles Whitman, an engineering student at the University of Texas, Austin. The former U.S. Marine sharpshooter climbed to the observatory deck of the clock tower with bags full of survival gear, ammunition, and an arsenal of weapons, including handguns, a shotgun, and a high-powered rifle. Once there, he began firing at random passersby from above. He was eventually shot dead by police officers, but by then, 16 people were already murdered.

Prior to the killings, Whitman was seeing a handful of physicians for his chief complaint of recurring headaches. One of them, a psychiatrist named M. D. Heatly, recorded evidence of new and unusual volatility in Whitman's personality in his case notes:

- "Oozing with hostility."
- "Admits having overwhelming periods of hostility with a very minimum of provocation."
- "He recognizes, or rather feels, that he is not achieving in his work at the level of which he is capable and this is very disconcerting to him."
- "The youth could talk for long periods of time and develop overt hostility while talking, and then during the same narration may show signs of weeping."

Nothing in these sessions explicitly answers the why behind his murderous rampage. But excerpts from Whitman's suicide notebooks suggested that his heinous actions may have been more just than a random incident. He writes, "I have been a victim of many unusual and irrational thoughts" and "I do not really understand myself these days." He even suggests a reasonable course of action considering the changes to his mental state: "After my death I wish that an autopsy would be performed on me to see if there is any visible physical disorder."

This wish was granted. Soon after his body was recovered, the governor of Texas established a task force to try to make sense of the tragedy, starting with an autopsy. Much of the report describes the injuries Whitman sustained during the gunfight that ended his killing spree. But the most surprising find was a tumor, two centimeters (0.8 inch) long, adjacent to his amygdala, later discovered to be a glioblastoma, an extremely aggressive brain tumor. Naturally, controversy arose over whether this tumor played any role in driving him toward senseless acts of violence. I don't believe anyone thinks that his brain tumor caused him to start killing his loved ones and strangers. But a new change in behavior could be at least in part explained by these dramatic changes in his brain.

Since Whitman, more case studies have demonstrated the link between brain tumors and antisocial behavior. Surgery worked wonders for one pediatric patient, a 10-year-old girl with a long history of seizures that became progressively more severe. Her behaviors included inappropriate sexual grabbing, thoughts of killing her family members, and thoroughly documented coprolalia more befitting of a movie script for Samuel L. Jackson than of a fifth grader, including phrases such as "I'm sorry I said the F word, you fucking idiot." One year after surgical removal of a temporal lobe ganglioglioma, her seizures disappeared along with her inappropriate behaviors and R-rated language. This case study is not an exception: according to one estimation, more than 20 percent of patients report personality changes, whether it be new-onset anger, socially inappropriate behavior, or verbal aggression, before the detection of their brain tumors.

There was also the 1991 case of Herbert Weinstein. Police were called in to investigate the apparent suicide of his wife Barbara, who had fallen from the twelfth-story window of their Manhattan apartment. However, the case took a turn for the sinister on interviewing Herbert. The 65-year-old retired advertising executive with no criminal record or history of violence admitted to strangling his wife before throwing her out of the window. He showed no remorse whatsoever. After his arrest, brain scans revealed not a tumor but a massive arachnoid cyst, an empty space within the meninges about the size of an orange, compressing his frontal lobe. The prosecutor for the state of New York recognized that a jury would see the abnormal deformity inside Weinstein's head and would have reservations about whether he was in a state of mind to commit the crime. Instead of pressing for second-degree murder, Weinstein was charged with manslaughter, a crime that carries a lesser sentence. His defense argued that he lacked the responsibility and therefore the mental capacity for murder. He pleaded guilty, was sentenced to prison, and was released at the age of 79.

The introduction of brain scans into the realm of criminal law creates some very tricky issues for which there are not yet any clear answers. For example, in chronic traumatic encephalopathy (CTE), the brain undergoes dramatic shrinkage as molecular markers of dementia start to appear as a result of repeated head injuries. CTE patients exhibit impulsivity, suicidality, explosivity, and otherwise socially inappropriate behavioral changes. Could these changes predispose someone toward violence, such as the murder committed by former professional NFL player Aaron Hernandez? If so, how would brain scan evidence of a shrunken brain or psychiatrist testimony sway a jury toward finding someone not guilty by reason of insanity? In another example, consider John Hinckley Jr.'s trial for his attempted assassination of President Ronald Reagan. The court allowed presentation of CAT scan images that showed some brain atrophy consistent with schizophrenia, such as a widening of the sulci between cortical areas. However, this pattern of brain changes is not a surefire sign of the diagnosis. It appears in one-third of people with schizophrenia and in 2 percent of people without. Ultimately, the jury found him not guilty by reason of insanity.

Much to the dismay of *Minority Report*'s Precrime department or the phrenologists with their theorized "murder organ," there is still no basis for certain brain landmarks that can predict murder.

Is he man or monster or . . . is he both?*

The idea of one or more entirely different personalities sharing a single physical body has served as inspiration for so many figures in fiction across different media. Chuck Palahniuk envisioned a mild-mannered product recall specialist and the charismatic terrorist cult leader Tyler Durden (*Fight Club*). M. Night Shyamalan showed us a man with 24 personalities (*Split*), one of whom is a violent predator who kidnaps three girls in preparation for the arrival of another personality with supernatural powers. Marvel Comics has many lesser-known characters with multiple personalities, including Moon Knight (1975) and Multiple Man (1975). But the most well known is, of course, Bruce Banner, the genius nuclear scientist who transforms into the powerful Hulk (1962) when he's angry.

Although these characters were invented using tremendous creative liberties that exceed the accurate medical diagnosis, these fictional characters could be experiencing dissociative identity

* The question posed on the cover of Marvel Comics' debut issue of *The Incredible Hulk*, created by Stan Lee and Jack Kirby. Originally gray, modern fans are likely more familiar with the green-skinned giant whose purple pants are made of some impossibly elastic material.

disorder (DID), a condition that has previously been called multiple personality disorder. According to the diagnostic guidelines, a person with DID experiences a fracturing of their identity into two or more personality states. Like a confused Bruce Banner standing in a crater of rubble, they experience a loss of agency, memory, and consciousness when one of the other personalities sits in the pilot's seat. These other identities, informally called "alters," may have dramatically different ideologies, goals, and standards of morality. Some reports of DID suggest that the personalities even have different physical traits, such as hand dominance or the ability to see, the latter even backed by visual cue evoked potentials as measured by EEG. Some of these alters may take on different roles for the person, such as a caregiver who watches over the body to minimize risk, a host who is the personality most often at the front, or a persecutor who might be modeled after an abuser. The alters may be of a different age or gender, and some may not even be human.

For most psychiatric illnesses, the debate is typically centered around the etiology ("Does heavy cannabis use predict the onset of schizophrenia? Maybe, but it really depends on the population being studied.") or prevalence ("How do we control for external variables when considering how many people are affected by the disease? Prevalence of depression more than tripled to 12.3 percent in the years following the 2009 financial crisis in Greece.") But for DID, the debate is whether it even exists at all. Psychiatrists on both sides of the debate have strong words about their stance, as evidenced by the titles of academic publications defending their claims, including the following:

- "Dissociation Debates: Everything You Know Is Wrong"
- "Errors of Logic and Scholarship concerning Dissociative Identity Disorder"
- "The Persistence of Folly"

- "Disinformation about Dissociation"
- "The Rise and Fall of Dissociative Identity Disorder" and its counter publication, "The Rise and Persistence of Dissociative Identity Disorder"

Advocates for the existence of DID stand by the trauma model, which posits that DID is a protective defense mechanism. According to this framework, the patient experiences severe or prolonged abuse early in their childhood, often at the hands of a caretaker or parental figure. In response to that, they form a separate identity to isolate the experience, essentially compartmentalizing the psychological damage to just one personality to spare the other personality, allowing them to live a normal, healthy life. In support of this theory, many of the symptoms of DID resemble those seen in PTSD, such as emotional numbing and derealization. Furthermore, up to 90 percent of people with DID have a history of childhood abuse, lending further evidence to the origin of the disorder.

Opponents, however, suggest the socio-cognitive model, where the presentation of DID is a result of external influences rather than a genuine mental disorder. They point to observations such as the increasing number of DID reports that track with media depictions of the disorder. The number of diagnosed cases before the 1970s were in the hundreds, but this number increased between the 1970s and 1990s to the tens of thousands, they would suggest in response to the best-selling 1973 book *Sybil* and its 1976 made-for-TV movie starring Sally Field. This increase could be a result of patients who have been made more aware of their symptoms, or it may be an error on the part of the psychiatrists who might be eager to overinterpret symptoms. Therapists are in a position to present leading questions during patient interviews, perhaps something as blatant as "Is there a part of you that I haven't spoken with?" or something more subtle, such as hinting about an "inner child" that a fantasy-prone patient might later interpret as a younger alter.

Another aspect in the debate over the existence of DID is the number of false diagnoses made throughout the years, such as that of Shirley Mason, aka Sybil, about whom the 1973 book was written. According to Cornelia Wilbur, the psychiatrist who made the diagnosis, she had 16 different personalities, including Marcia, a painter who spoke with a posh British accent, and Mike, a carpenter who was preoccupied with a desire to "give a girl a baby." Wilbur was a strong proponent of the trauma model as an explanation for the development of Mason's symptoms, using her background as a Freudian psychiatrist to suggest that repression of forbidden sexual desires drove the genesis of the dissociation rather than genuine trauma. Unfortunately for those psychiatrists seeking evidence for the authenticity of DID, Mason's case represents one point scored for the skeptics. Mason had admitted to another psychiatrist that she did not actually exhibit the characteristics of multiple personalities but rather had been guided to believe that under the influence of sodium pentothal and hypnosis. In an even more heinous violation of professional ethics, Wilbur had a book deal to publish Mason's story, providing her with an incentive to diagnose the fascinating condition for personal financial gain. At the root of the controversy is Sybil, Incorporated, a business entity created for the purpose of sharing revenue from the frankly bizarre merchandise, including T-shirts, a jigsaw puzzle, and even board games.

Diagnosis of DID still relies on subjective reporting from both the therapist and the patient. Demonstration of some cut-and-dry biomarker found only in patients with DID would represent a major step toward ending the debate and establishing the existence of this tricky diagnosis. No such evidence has been found thus far. Anatomically, these patients have a significantly smaller hippocampus and amygdala. But this pattern is also seen in PTSD and major depressive disorder, making it a nonspecific observation. As for measures of functional connectivity, one model of psychopathology argues that some differences across three major neural networks may explain a variety of psychiatric illnesses. One of these networks, the salience network, is involved in the switching of internal and external stimuli. People with DID show less connectivity between this network and the other networks. This may account for why the source of some stimuli may be confusing as to where it comes from, which is then interpreted as being understood by only one alter but not another.

The neuroscience questions surrounding DID are interesting but not nearly as fascinating or as difficult, as the legal ramifications if DID is a widely accepted diagnosis. It has already been used in criminal court successfully for one defendant and unsuccessfully for at least another. In 1978, Billy Milligan was on trial for multiple armed robberies, kidnapping, and rapes based around the Ohio State University campus. Before his trial, his psychological evaluation revealed DID with a total of 24 unique personalities. According to the psychiatrist, Ragen, a 23-year-old Yugoslavian, had committed the robberies, while Adalana, a teenage lesbian, had committed the rapes. At his trial, his defense argued that Billy and other personalities were not responsible for Ragen's or Adalana's crimes, so any punishment appropriate for the guilty would be also unfairly assigned to the nonguilty personalities. The jury found him not guilty by reason of insanity. Billy was acquitted for the crimes and sentenced to confinement at psychiatric hospitals for treatment.* On the other hand, the DID defense didn't work for Kenneth Bianchi, the man convicted of several murders around the Los Angeles area. He had blamed an alter named Steve for the murders, although psychiatrists saw through the malingering during his 1979 trial. In response to being told that patients usually have more than two alters, he produced new personalities on

* A medicolegal description of the case, *The Minds of Billy Milligan*, was written by Daniel Keyes, the author of *Flowers for Algernon*.

demand. There was also a lack of consistency of personality traits of each alter, and no one close to him had ever described a sudden dramatic change in personality, indicating that his claim of DID was likely fabricated for the sake of a less strict sentence.*

From the legal perspective, patients with DID share similarities with homicidal sleepwalkers. A person who is not in control of their body at the moment when a crime is committed may also not be able to process that crime and therefore may not be held liable for those actions. But from the jury's perspective, it is far easier to believe the authenticity of a condition that presents with visible evidence, such as a defendant's nighttime EEG brain activity bouncing abruptly between the phases of sleep. It is still very difficult to provide irrefutable laboratory evidence of DID.

> I think anybody who falls in love is a freak. It's a crazy thing to do. It's kind of like a form of socially acceptable insanity. —*Her*

Most biologists can effortlessly write about aromatase inhibitors or the anterior cingulate gyrus in their publications, confident they are able to describe them in concrete terms of mechanistic action, anatomical location, or a functional effect of these things on animal behavior. There is one concept that many are frankly scared to write about, a behavior that is found across cultures globally, likely conserved over millions of years of hominid evolution and possibly the only reason any of us are even here. And if you could calculate the strength of this one behavior by adding up the number of songs, movies, paintings, plays, poems, architectural wonders, and other works of art it has inspired, it is more influential than any other force in the universe.

Love.

* In support of the socio-cognitive model of DID, both of these cases were in the years shortly after *Sybil*.

Many scientists tiptoe around the indefinable concept of love by substituting sterile phrases, such as Darwin's "mate choice," "sexual favoritism," or the comically devoid of emotion "selective proceptivity." But not Helen Fisher. Considering Fisher's appointment at the Kinsey Institute, the most famous sexual health research institute in the world, it is no surprise that she uses the word liberally in her academic publications. According to her most famous theory, long-lasting romantic love is a three-step process, each characterized by different behavioral patterns. First is lust, that animalistic instinct that drives mating. Lust is fired up in seconds and is thought of as an immediate-term response fixated on physical features and sexual gratification. Second is attraction, a phase where every thought is preoccupied with that special person. Third is attachment, the long-term phase where plans are made on the timescale of years, decades, and generations.

Considering that each of these phases is characterized by different behaviors and accompanying personality traits that change as love evolves, there is no surprise that different underlying biological processes are involved with each. Fisher's theory accounts for these processes, offering up some of the signaling molecules involved with each phase. Lust is driven largely by neurohormones such as testosterone and estrogen, synthesized by the sex organs, and then dumped into the bloodstream. The former is notorious for increasing impulsivity and risk-seeking behaviors in men and women alike, causing us to cast the widest possible net whenever searching for a partner.* Attraction is driven by dopamine and serotonin. Midbrain dopamine is part of the reward pathway, spiking with every notification on your phone in anticipation of hearing from your partner. As for the last step, attachment involves oxytocin and vasopressin, two molecules synthesized by the hypo-

* To use an analogy more relevant for those of us who don't fish, the sex drive causes us to swipe right on everyone.

thalamus and then released into the bloodstream by way of the pituitary gland. Attachment offers security and safety, things that are missing in the risky, uncertain phases that precede it.

Surprisingly, there are nonhuman models used in testing questions about romantic interactions. One such model is a species of fuzzy little rodents called prairie voles, commonly found across central North America. Many consider them as pests for their habits of damaging root systems and tearing up gardens. But neuroscientists find them fascinating for their socially monogamous lifestyle, a behavioral trait that is seen in only 9 percent of mammalian species.* Once a pair of wild voles mate, they spend much more of their time with their partner, living together in their underground burrows, sharing the food they find after expeditionary trips, and taking turns caring for their young. This pair bond lasts until the partner vole dies: 'til death do they part.

A shift of behavior into a lovestruck, caring, homebody can be demonstrated with a set of cleverly devised laboratory experiments. In the partner preference test, the male of one mated pair is tethered to a chamber. An unfamiliar male is tethered to another chamber. Then the female vole is placed in a neutral chamber and given the option to go visit either her partner or the new mystery vole. Just like an infatuated human who craves her partner's company, she will spend more time with her paired couple over the stranger. She also spends more time huddling with her partner, lying down side by side with him. If you separate a pair of mated voles from one another, they exhibit traits of rodent anxiety. If you unexpectedly introduce a new male vole into the happy little home of the "newlyweds," the original male quickly turns into a violent aggressor, biting and clawing at the intruder in a bout of jealousy.

With an experimental model in place, scientists began meddling with prairie vole long-term relationships. Through this

* But nearly 90 percent of avian species mate for life. Monogamy is for the birds.

interference, they found that oxytocin is strongly involved in this behavioral shift. If a female prairie vole is given an antagonist for the oxytocin receptor during their more intimate moments, thereby nullifying the effect of any oxytocin released, she fails to show any preference for her partner in the future. Conversely, a female vole that cohabitates with a male sans sex while receiving oxytocin behaves as if she had mated with him. She spends more time with him over random voles if given the choice, she nuzzles up against him longer, and she acts heartbroken when separated from her chemically induced beloved.

When Cupid takes aim at a prairie vole, he dips his arrows in oxytocin.

An advantage of studying the prairie voles is that there is a phylogenetically similar animal, the montane vole, that shares physiological features but does not pair bond after mating. Male montane voles mate indiscriminately with any female montane vole in estrus. By the time the litter is born, the father voles have long moved on to the next female. And if you put them in the partner preference test, they spend as much time with the stranger vole as they do with their mated partner. These make for an excellent comparison population to better understand the differences between monogamous and polygamous voles. In comparison, prairie voles have dense expression of the oxytocin receptor in their nucleus accumbens, a structure that contains several dopaminergic terminals that are involved in reward. This pattern of expression at its simplest implies that activation of these circuits through monogamy encourages these types of actions further in a feedback loop.

Despite the totally coincidental observation that "vole" is an anagram for "love," this experimental model still doesn't approach the nuanced complexity of human relationships. Vole researchers are particularly cautious about using the phrase "pair bonding" instead of love. Love takes on many forms, and no one wants to believe that the exhilaration of spinning madly out of control is just a result of molecules. Fisher herself readily admits that her "three-phase" framework is a simplification of the way love works in reality. For one, it minimizes the influence of cultural and societal standards. Consider consensual arranged marriages, such as those seen in India. Here, long-term stability is prioritized, which mitigates some of the risk involved in potential partners. It also doesn't account for changing trends in relationships, such as how technology has increased interpersonal connectivity, making meeting potential partners ironically both easier and more difficult.

> I'm not alive! You're not going to bed with a woman. Don't be disappointed; okay? Have you ever made love to an android before? —Philip K. Dick, *Do Androids Dream of Electric Sheep?*

Your heart starts beating pounding, practically thumping out of your chest. Your breathing quickens to keep up. Your skin is flushed, and your armpits feel damp. Was this the body's natural stress and fear reaction precipitated by the thought of hunting down a ruthless leader of murderous androids?

Or did you just meet the love of your life?

The two very different emotions of sudden fright and romantic love produce the same bodily physiological changes. And according to one theory based on this observation, people conflate fear with love, a confusion known as misattribution of arousal. To demonstrate this phenomenon, researchers sought a way to reliably evoke a sensation of mild fear in a naturalistic setting, somewhere out in the field where they would recruit unknowing participants for their study. The Capilano Suspension Bridge in North Vancouver, British Columbia, was the perfect spot. At 450 feet (127 meters) long, it sags a little bit in the middle, weighed down by its own cedar planks and steel cable connectors. A breeze sets the bridge swaying just a tiny bit, exactly as it was engineered to do, but that movement can be disorienting, especially if you are looking down. The mesh fencing will keep your foot from slipping off the path, but it doesn't keep you from imagining what it would feel like to free-fall 230 feet (70 meters) into the river or onto the rocks below.

In their experiment, "an attractive female interviewer"* was positioned at the middle of the bridge, waiting for potential research participants to cross. She would occasionally stop young men without an apparent companion and ask them if they were willing to help her out for a psychology class. She would explain that she is studying the effect of scenic views on creativity. To measure creativity, participants were shown a vague picture: that of a woman coyly covering part of her face with one hand while extending out the other. Participants were instructed to write a short background story that provides context to the picture. At the conclusion of the experiment, the woman handed the participant her name and number and invited them to call her to learn more about the results of the experiment. Other control experiments were also done, including recruiting a male interviewer and stopping people over a control bridge made of solid cedar only 10 feet above a trickling rivulet that served as the low-fear condition.

A little deception is an expected part of many psychology studies: the researchers didn't actually care about creativity. They were curious how much romantic or sexual content was involved in these fictional backstories. The men crossing the nervousness-provoking experimental bridge were significantly more likely to include some description of a girlfriend, lover, kissing, or a marriage proposal compared to the much more mundane answers offered by men crossing the safer condition ("I think this woman is a little embarrassed about something that is not important at

* Yes, this subjective judgment on the attractiveness of their coworker/student was explicitly stated in their publication. The 1970s was a weird time.

all."). According to the authors, the increased romantic content was a consequence of putting these men into a state of apprehension or nervousness, one that physiologically mimics the sensation of love, and this high-arousal state translated into the context they created for the image. Furthermore, the participants who crossed the scary bridge were significantly more likely to "[phone] a strange female to effect a liaison"—they were given the lab phone number—compared to those who crossed the control bridge (50 percent vs. 13 percent). Scared men are more romantic men, and more likely to take risks for love.

Misattribution of arousal might explain how Kristin Enmark felt when her workday was disrupted one day in Stockholm, Sweden. She was certainly scared for her life when the gunman, Jan-Erik Olsson, entered the bank wearing a wig and makeup and then fired his submachine gun wildly into the ceiling. He demanded that his friend be released from prison and transported to the bank, where the two of them were to be given a stack of money, weapons, armor, and a getaway vehicle. The plan was to take the hostages—Enmark and three other bank employees—together in the car and leave the scene. The ransom demands were met quickly, but the authorities firmly denied their request to take the hostages with them as they escaped. In response, the two criminals barricaded themselves along with the hostages in the bank's safe. Olsson made it clear he was not afraid to use violence. He shot one of the officers who responded to the bank's silent alarm. When the police tried to intervene by drilling a hole in the vault to spy on the situation, he recklessly discharged his firearm into that hole. He also threatened to shoot his hostages.

For the next several days, life was tense outside the bank. The police had set up sniper posts covering every entrance and exit of the building. At peak media coverage, more than 70 percent of the Swedish public was tuned in to the live television feed, watching the doors closely in case something dramatic were to happen. However, inside the bank, the feelings of fear began to shift. After being held in the vault for hours, the hostages started to develop a rapport with the robbers, something less like fear and more akin to long-term trust. At one point, a hostage even spoke with the prime minister of Sweden on the phone, begging him to allow the two criminals to go free with hostages in tow. Enmark trusted the robbers more than the police, even expressing a fear that the police officers would overreact in the situation and accidentally kill the hostages. The hostages started feeling closer with their captors than with those trying to rescue them. In the words of one hostage, "This is our world now. Whoever threatens this world is our enemy." Tear gas was pumped into the bank, and when the 130-hour-long ordeal was over, the hostages were released, and the duo of criminals were taken into custody. Enmark called out to the gunman, "Clark, I'll see you again." None of the four testified against Olsson; on the contrary, they had raised money for his defense.

The connection between Enmark and her captor is believed to be an example of Stockholm syndrome, the controversial observation where a person taken hostage and therefore in a high-fear state begins to develop a close emotional bond akin to romantic love with their captors. The term appears much more in common parlance than in the medical literature, especially because there are no concrete diagnostic criteria for Stockholm syndrome in any classification system for mental disorders. Psychologists suggest that this attachment is a defense mechanism to improve the likelihood of sheer physical survival in the face of learned helplessness, even to the detriment of psychological well-being. It is possible that Stockholm syndrome is much more common than when people get held hostage: it might be a contributing factor as to why people have a difficult time leaving abusive relationships.

Get away from her, you bitch! —*Aliens**

Romantic love sometimes lays the foundation for a different type of interpersonal bond, that of parental love. And evolutionary biologists have a controversial take about this field.

Marching under a banner decorated with the face of their champion, Richard Dawkins, some have no interest in discussing the intangible and therefore ignore anything involving deities, intent, values, and other malleable concepts, such as right and wrong. Instead, they look only at the concretely quantifiable. They care about body shape, animal behavior, and, most recently of all with the advent of molecular biology, genes. The most hard-line faction within this group takes the extreme reductionist interpretation of life, suggesting that our bodies simply exist as the most efficient possible tool for maintaining the existence of certain sets of DNA molecules over the decades, centuries, millennia, and beyond. Millions of years of evolution would have fine-tuned every element of our being for just this purpose, such as the explosion of sex drive in the teenage years and labor pain amnesia, just so that those molecules can exist.† DNA is the pilot; we are the oversized bags of meat built for sex.

In support of this simplification of our existence, there are a handful of behavioral changes in the face of becoming a mother that are driven in part by the actions of oxytocin, the same molecule proposed in Fisher's third phase of romantic love. It smoothly

* *Aliens* plays heavily into themes of motherhood. In the final moments, protagonist Ripley defends a young girl named Newt, for whom she has formed a protective mother–daughter bond with. The body horror underlying the original *Alien* was male impregnation, as the protagonist was scripted to be a man. H. R. Giger, the artist who conceptualized the xenomorphs, incorporated phallic elements into the design, including their eggplant-shaped heads and their dependence on oral penetration for egg laying and reproduction.

† However, evolution still hasn't solved the fact that giving birth is still one of the more dangerous activities for many mothers even though some risks are mitigated by the tools of modern medicine.

bridges the narrow distance between romantic attachment to a partner and caring attachment to a baby.

The birth of oxytocin research begins with, well, birth. In 1906, Sir Henry Dale blended up some pituitary gland and injected it into the bloodstream of a cat. In response, there was a rapid contraction of uterine muscles, prompting him to name whatever chemical was here oxytocin, derived from the Greek words meaning "swift birth." Oxytocin's involvement continues after birth in lactation, one of the behaviors that defines what it means to be a mammal. Circulating oxytocin activates receptors found in the mammary glands, which results in contraction, allowing for the milk ejection. In a testament to the creativity of evolution, the letdown of milk is part of a positive feedback loop between the somatosensory nerve endings of the nipples and the hypothalamus. When newborns suckle, they stimulate the sensory component, which initiates further milk letdown, thereby producing more milk. There are also auditory circuits that project through the thalamus before reaching the hypothalamus. This neural circuitry explains why hearing the cries of an offspring can initiate milk letdown.

Beyond physiology, oxytocin is also involved in the complex set of changes to the psychology of motherly personality traits. A group of women were put into an fMRI after receiving a nasal spray of either oxytocin or a control saline spray. They were given headphones through which the sounds of crying babies or control sounds were played. For the saline-exposed people, the baby cries caused a huge increase in amygdala activity, that structure involved in stress and the sympathetic nervous system activity. But for those who received the oxytocin spray, that response was dampened. Furthermore, in the oxytocin condition, the signal in the insula, a structure containing circuits involved in empathy. increased. Collectively, this suggests that oxytocin promotes good motherhood: better management of her own stress in response to the baby's distress and a better understanding of the baby's emotions means she is more attentive toward her young.

The involvement of oxytocin on parenting behaviors extends even into early childhood. Mothers and their four- to six-year-old children were put into a playroom furnished with all sorts of delightful distractions, such as an Etch-a-Sketch, some drawing utensils, and cartoons looping on a television screen. There were also a bunch of unfinished chores, such as scattered clothes and crumpled pieces of paper. The mothers were tasked with asking their child to complete a certain number of tasks in order, including one involving some self-play, while she took an "important phone call." The mothers were then judged on how often they used praise and positive reinforcement compared to negative commands and criticism. They were also put into an fMRI and shown pictures of their children and demographically matched children. Using a genetic analysis, one specific variant of the oxytocin receptor was correlated with greater demonstration of positive parenting traits and fMRI signal across the anterior cingulate cortex, suggesting that oxytocin can drive better regulation of their own emotional state and stronger empathic understanding of their child's needs.

Although there are clear benefits of oxytocin in the context of motherhood, these boons may not extend to social interactions outside that realm. There are subtle effects in the literature showing gender-dependent differences in specific tasks, some of them hinting at a prosociality effect in autistic people and some reporting no effect, meaning that oxytocin's effect is below the threshold of what is measurable by our current in vivo human imaging tools. But shaky evidence isn't going to keep various internet retailers from trying to market the drug, many of whom suggest to mist a little bit on your clothes before social interactions. They offer such wild promises as "create a feeling of trust" or "keep relationship strong" and are even advertised as being "great for salesmen." Despite the assurance of chemical purity, dissatisfied customers are prone to leave scathing one-star reviews, such as "I don't think this works I didn't feel anything" and "No effect whatsoever."

In retrospect, I can't help but wonder how many people stumbled on this product while searching for oxycodone.

CHAPTER TEN

"ARE WE NOT MEN?"*

Who Is an Us, and Who Is a Them?

For storytellers, there is a shortcut for cultivating fear or distrust in an antagonist. First, appeal to humanity's collective will to survive. Then introduce an Evil, which also wants to survive. Finally, provide instances of how their survival runs in direct opposition to ours.

Take the pod people in Jack Finney's *The Body Snatchers* (1954) or its many adaptations over the years. The spores of these plant-like beings drift across space, eventually landing in the small suburban town of Santa Mira, California. These spores mature into pods, each one containing a near-perfect replica of a human, complete with memories and personality traits but lacking emotion. If unchecked, these aliens can deplete entire planets of resources, leaving Earth an empty husk when they float away to the next solar system. If it weren't for their collective decision to leave, it wouldn't be possible for these aliens to land on our planet and cohabitate peacefully with *Homo sapiens*.

A more subtle approach to creating a conflict is to find ways to differentiate the heroes from the villains. Traditionally, portraying these two opposite forces was easily achieved. Simply give your characters ties to culturally shared knowledge. For the Western

* An often-repeated phrase in the set of laws developed by the humanimals in H. G. Wells's *The Island of Doctor Moreau* (1896).

literary world, that meant the Christian Bible. Bram Stoker's *Dracula* (1897) is practically dripping with Christian lore. Count Dracula, whose name is roughly translated as "dragon," clearly serves as a stand-in for biblical evil: translations of the book of Revelation use the word "dragon" to explicitly refer to Satan. This is foreshadowed early on, when protagonist Jonathan Harker is given a crucifix and encounters throngs of people making the sign of the cross to bless him against something. Dracula stands in defiance of death itself; he speaks of the events of the past centuries using present-tense verbs. For Christians, resurrection is a holy miracle and an act of God (see Lazarus, Jesus, book of Revelation), so for a being to overcome death is to defy God's will. Dracula's thirst for human blood is a perversion of the role of blood in the sacrament of the Eucharist. Dracula is weakened by sacred communion wafers, and his final demise is brought about by a wooden stake through the heart, a twist on the crucifixion of Jesus. Stoker, an Irish Protestant himself, clearly planned out his depiction of Dracula as an Antichrist figure.

Beyond *Dracula*, there is a whole genre of modern religious horror that calls on the collective shared knowledge of Christianity. The most culturally significant of these may have been *The Exorcist* (1973), which depicts the possession of a young teenage girl and the priests who fight to release her from demonic influence. The film set all kinds of records across America. People were lining up at 4:30 in the morning to get tickets for a 10:30 show. Law enforcement was sometimes brought in to keep order over these lines, which stretched around the block. Theatergoers were fainting and seizing. Sales of pea soup plummeted.* A psychiatrist even published a case report in the *Journal of Nervous and Mental Disease* of four patients who experienced "cinematic neurosis," char-

* I don't know if this is true, but I also don't know if it's *untrue*.

acterized by recurring flashbacks of actress Linda Blair's swiveling head, debilitating anxiety over becoming possessed, and substance misuse to eliminate memory of the film. *The Exorcist* is just one of the many examples of horror media that incorporate elements of Christianity to orient good from evil; others include classics such as *Carrie* (1974) and *Hellraiser* (1987)* and more modern works such as *Midnight Mass* (2012) and *The Nun* (2018). Some of these put a creative twist on the trope, painting fanaticism as the real evil.

Religion is just one thread of the tapestry of culture. From entertainment and art to economic structures and interpersonal relationships, human societies differ from each other in myriad ways. Tiny linguistic differences, even among speakers of the same language and within the same country, can produce some laughable moments, such as when ordering a coke at a restaurant in the American South, where "coke" is used generally to refer to any soft drink. Differences in holidays and celebrations sometimes produce mild confusion, such as when overhearing a group of Canadians speaking about preparing for Thanksgiving in the middle of October. Some traditions provoke a sense of moral outrage. For the Kaluli people, young boys receive semen through the anus from an older man as part of a rite of passage into adulthood, an act that the nearby Etoro people find repulsive—they believe the semen must be swallowed. Even entertainment has led directly to violence. After the Port Said riot in 2012, 74 Egyptians were dead, more than 450 were injured, and another 21 were found guilty in the court of law and sentenced to death. The riot broke out at a football match.

With a little bit of creativity, any one of these elements can be leveraged to create a sense of paranoia against any group that

* Technically, Clive Barker's Cenobites from *The Hellbound Heart*, the novella that inspired the *Hellraiser* franchise, aren't explicitly from hell. To quote the haunted-house-in-space film *Event Horizon*, "Hell is only a word."

doesn't share the same cultural values as "us." Authors and screenwriters use othering whenever they want to remind us of who we want to cheer for. Whereas this brand of othering had its place in the arts, it has serious implications in the real world. It doesn't take a lot of work to instill a sense of mistrust against people who are different from us. Long ago in evolutionary time, there was likely a survival benefit in being able to judge in-group from out-group. Some people would be helpful, trustworthy, and cooperative, offering food or shelter to your protohuman ancestors in need. There would be an expectation of reciprocation among the in-group. Members of the out-group, however, might take advantage of your kindness for their own gain. Worse yet, they might be outright hostile. Today, totalitarian and fascist regimes develop propaganda using othering whenever they want to vilify a society, marginalize a population, and rationalize genocide.

Behaviors conserved through evolution likely maintain traces of the neural mechanisms that explain said behaviors. In the previous chapter, much space was dedicated to extolling the many virtues of the hypothalamic neurohormone oxytocin. Yes, it plays a role in love, setting a solid foundation for long-term attachment and motherhood. But this view oversimplifies the truth of the hormone. Oxytocin has a dark side. It widens the metaphorical gap between the Us and Them.

Researchers recruited indigenous Dutch males and gave them a snort of either oxytocin or a placebo nasal spray. Participants were seated at a computer and then assessed on how they would act in the classical ethics dilemma the trolley problem. They were told to imagine that a runaway trolley is on course to kill five innocent people. To save those lives, they could simply flip a switch, which would divert the trolley onto a track where only one person will be killed. Similar variations were also presented, including one in which a man must be denied access to a lifeboat to ensure the survival of the others onboard and one especially gruesome scenario where a man is lodged in a cave tunnel and must be blown up so that those trapped behind him do not drown. To give the participants a more personal investment in the scenarios, researchers named the man who could be sacrificed. These names were picked strategically to instill an us-versus-them mentality. One set of names included Ahmed, Mohammed, and other stereotypical Arab names. At the time of the study, many Dutch citizens held generally unfavorable opinions against Muslims, likely a remnant of the spike in Islamophobia in the wake of 9/11. The second set of names were German names, such as Markus or Helmut. While ethnically more similar than Middle Easterners, some Dutch people nonetheless held negative stereotypes of Germans, who were perceived as being aggressive or arrogant. Finally, a set of in-group names served as the control, which included Dutch names, such as Dirk and Peter. While under the influence of increased oxytocin, Maarten was more likely to be saved, but innocent Youssef and Wilhelm were more likely to be sacrificed.

Oxytocin also affects performance on facial recognition tasks. Generally, we are better at recognizing faces of people who are of the same racial background compared to those of a different background, an observation called the other-race effect (ORE). The ORE has very real consequences. Faulty eyewitness testimony, already terribly flawed on its own, has turned innocent people into criminals. In fact, the ORE was first described more than a century ago in the context of law and criminology, suggesting that sex, race,* and age are the three key pieces of information typically used to describe a perpetrator. Whenever a person had to identify a person of a different race in a lineup, they performed poorly. Among convictions that have been overturned due to DNA evidence, nearly all of them were due to incorrect eyewitness testimony to

* The 1910s, when the article was published, was not known for nuance. The author suggested there were five races: "white, yellow, brown, red, or black."

begin with. A little snort of oxytocin decreases the intensity of the ORE, increasing performance on out-group recognition. An optimist might say oxytocin eliminates the "all look same" feeling you might get when seeing people of different races, having an altogether prosocial effect. But the pessimist suggests that oxytocin helps us better remember which one of those Others harmed us, allowing for targeted retaliation in the future.

Like other of its better-known neurochemical brethren, oxytocin has been given its own catchy rhyming phrase to broadly describe its influence. If norepinephrine signals "fight or flight" and acetylcholine encourages "rest and digest," oxytocin pushes "tend and defend." It helps us better care for those in the in-group, a task that is most easily accomplished if there is a hard line that separates those people from the out-group.

> After rushing to the station at 7:30 I had to sit in the carriage for more than an hour before we began to move. It seems to me that the further east you go the more unpunctual are the trains. What ought they to be in China?" —Bram Stoker, *Dracula*

To signal the Evil, you don't even need to show an interplanetary race bent on universal domination or a supernatural revenant seeking to avenge some ancestral wrong. You need only remind the audience that humans already have a difficult time coexisting peacefully with other humans.

Granted, there is one particular tradition that is taboo in practically every culture and is therefore a surefire way to signify a Them. The neuroscientific study of this practice starts with a strange disease that affected the Fore people, an indigenous tribe of Papua New Guinea. The victims started experiencing progressively worsening muscle control and coordination. They walked with short, stuttering steps, sometimes causing them to fall. Soon after, they would develop a nonstop tremor in their limbs and exhibit signs of emotional dysregulation. Then they would lose all control over bodily functions, completely unable to move. A year after the onset of symptoms, the patient was dead. Within the five years since its first documented appearance in 1957, more than 1,000 people died. Some called it the "laughing disease" due to the wild peals of hysterical laughter that came from the patients. However, the Fore people called it *kuru*, meaning "trembling" in their native language.

In 1961, a young Australian doctor named Michael Alpers, having recently graduated from medical school, chose this nontraditional assignment for his first job. He packed up and moved to the Eastern Highlands of Papua New Guinea, living among the Fore people and hoping to do something about their plight. Over time, he integrated with them. He learned the language, made friends, and became familiar with their customs and traditions. He teamed up and collaborated with the American doctor Daniel Carleton Gajdusek, who had also been studying the disease. He also saw a handful of the estimated 200 people die each year from the dreaded trembling disease. He also observed that *kuru* seemed to target a subpopulation within the tribe: mostly women and young children, while infected men made up only 2 percent of total cases. This discrepancy was so pronounced that in some villages, men outnumbered women by almost three to one.

At this point in the 1960s, no one knew how the disease was passed around. What was known is that in response to disease-causing agents such as a cold or a flu virus, the body recognizes these invaders and reacts accordingly by launching a defensive cellular assault against them. This process leaves a very long-lasting impression on the body, an "H1N1 was here" graffiti-like mark that can be picked up with a blood sample. But in the patients who died from *kuru*, no such tag was found. At a dead end, researchers shifted focus away from molecular markers and toward elements of Fore culture. Was there some toxin in their food? Samples of tree bark, leaves, stems, berries, and herbs that the locals ate came up negative for anything neurotoxic. But they were on the right track by looking at their diet.

Grieving the loss of a loved one is a universal human experience. For the living to better cope with this, every society has developed their own formalized farewells reflective of their values and traditions. For many cultures, these rituals revolve around wishing the best for the deceased by preparing them for the afterlife. Think of the 8,000 or so painstakingly sculpted terra-cotta soldiers buried alongside the first emperor of China or the massive stockpiles of food, jewelry, and furniture entombed with the pharaohs of Egypt.

The Fore people ate the deceased.

Soylent Green is people! —*Soylent Green**

According to the Fore, a body buried in the ground would become a meal for worms, while a body left aboveground would become food for flies. Neither outcome was a pleasant thought for the survivors. By engaging in ritualistic cannibalism, a part of the body would remain immortalized within the living relatives of the deceased. Eating the body was seen as an expression of grief and a way to cope with loss.

The usual suspects of disease-causing agents, such as bacteria, viruses, fungi, and parasites, didn't fit the mold of *kuru*. It was nothing like anything previously described that caused illness. Instead, the offending particles were proteinaceous infectious agents, or prions for short. Prions seem to ignore the muscles and skin, residing mainly in nervous system tissue, especially in the brain and

* The 1973 film *Soylent Green*, inspired by Harry Harrison's novel *Make Room! Make Room!*, depicted a dystopian future where the Soylent Corporation solves Earth's two biggest problems simultaneously: overpopulation and world hunger.

spinal cord. During the feast of transumption, the body parts of the deceased were separated, and each part was eaten for a specific purpose and by a specific recipient. The muscle tissue was reserved mostly for the men, who believed that eating this tissue would improve their skill in combat. Women, however, preferentially ate the brains. The Fore believed that everyone had five different souls. One was called the *kwela*, and it represented the pollution of the flesh. If the *kwela* was not properly sequestered, then it would haunt the living descendants, and the only force powerful enough to tame the *kwela* is a woman's womb. Also, women were almost exclusively responsible for the preparation of the body. Without washing themselves after the gruesome dissection, they might feed their children or offer up a small bite of brain as a snack to their children. This explained the disparity in who became infected. Soon after the epidemiological and biochemical studies had been completed, the Papua New Guinea government outlawed the practice of funerary cannibalism. The Fore stopped the tradition, and within a generation, the disease was essentially eradicated. Two Nobel Prizes in Physiology or Medicine were awarded because of *kuru*. One was given to Gajdusek* for his work understanding the epidemiology of the disease and the other to biochemist Stanley B. Prusiner for his molecular characterization of prions.

The knowledge gained from prion research has saved many other lives, especially during the outbreak of mad cow disease in the 1990s. The affected animals experienced a loss of muscle control and unusual aggressiveness or anxiety. The cause was tied to a disgusting infrastructure that relied on cannibalism. Many slaughterhouses were feeding healthy, uninfected cows with a processed form of cow carcass that was contaminated with diseased nervous

* His extensive journals describe much more than his *kuru* research. They also detail his pedophilia toward young boys of the places he visited. After admitting to sexually abusing one of the many boys he brought to live with him in the United States, he served a year in prison before spending the rest of life in exile in Europe.

tissue. Millions of cattle were slaughtered and their bodies discarded out of risk of having been exposed to prions. The beef trade slowed in response to cases when the infection jumped species to humans, resulting in the neurodegenerative prion disease variant Creutzfeld-Jakob disease (vCJD). The symptoms in humans are like those seen in cows and are equally fatal, patients typically dying a year after presentation of symptoms. Understanding how the disease is transmitted among livestock helped us develop various precautionary measures. Carefully separating out the brain and spinal cord during slaughter was one safeguard against the further spread of disease. And of course, stopping the cannibalistic use of deceased animals for feed further minimized the risk of another mad cow disease outbreak. The human vCJD has been on the decline since 2000. And, for the Fore people of New Zealand, they are no longer afflicted by the terrible laughing disease *kuru*: the country has been *kuru* free since 2005.

> The throngs of people that seethed through the flume-like streets were squat, swarthy strangers with hardened faces and narrow eyes, shrewd strangers without dreams and without kinship to the scenes about them, who could never mean aught to a blue-eyed man of the old folk, with the love of fair green lanes and white New England village steeples in his heart. —H. P. Lovecraft, *He**

If there is one piece of trivia about H. P. Lovecraft that the non-horror aficionado knows, it is his views on race. His ethnocentrism (how dare these immigrants pollute my ears with their gibbering, blasphemous languages! Gasp!) and white supremacist ideologies (those slant-eyed Orientals, up to no good) fueled much of his

* *He* tells the story of a man who moves to New York and meets someone from the long past who reveals the terrible dark secrets of the town's history. Written in 1926, *He* was published shortly after Lovecraft's brief stint in the big city. Lovecraft does little to disguise his opinions about Greenwich and its cosmopolitan residents.

writing. He broadly painted the villains as being dark-skinned or adherents of non-Christian religions. A fear of miscegenation pops up in his work, from the horrors of the human residents of Innsmouth, who interbreed with the frog-like Deep Ones, to his protagonist Arthur Jermyn, who discovers that he was the descendant of an African ape goddess–human union—to which his reaction is a dramatic self-immolation in the middle of the night. In addition, there's the name of his real-life cat.*

Modern racism, while less overt than in the early 1900s, still haunts minoritized people today. It is no surprise that themes of race continue to appear in today's horror and science fiction to reflect the authors' experiences. *Get Out* (2017) was comedian Jordan Peele's entry into the genre. In it, a Black American meets the family of his White girlfriend, who develops a surgical process to transplant the consciousness of a person into a new body, leaving the host consciousness trapped in a hypnotized comatose state called the Sunken Place. Between all this mad science and the other strange things that happen throughout the front 99 percent of the movie, it is Peele's depiction of the state of race relations for Black Americans in the final three minutes that produces one of the most cinematically tense moments in modern film.†

Many cultural practices can be disguised. No one knows which deity you address your prayers toward, and a cannibal can go several years without eating a single spoonful of brain. But some physical differences cannot be hidden. Short of plastic surgery or identity scrambling suits such as those in *A Scanner Darkly*, you can't hide your skin tone or the shape of your epicanthic folds. Calling attention to these physical traits is the shallowest form of othering but one that has been used to justify a hierarchy of characteristics, laying the foundations for many evil deeds.

---

* It's not my place. Look this one up online.

† After the protagonist fights through an underground research facility, he escapes to the outside and strangles his body-snatching girlfriend on the road. Right at that moment when we think he is nearly free, red and blue police lights flash behind him.

The idea of a race being encoded in genes is largely pseudoscientific, as there is no biological definition of race. In fact, there is more genetic variation within racial groups than between groups. Regardless, that hasn't stopped people from trying to distort science to justify being horrible to each other. Phrenology was flawed in a bunch of ways, but its frequent reliance on stereotypes had many of its roots in racism. Some diagrams depicted people with a prominent aquiline nose, a trait commonly used in an anti-Semitic context (although it also frequently appears among South Asians, Native Americans, and people of Mediterranean descent, hence why it is also called the Roman nose). Of course, the drawings of these people were given negative labels: unreliable husband, an all-around "weakness of good traits," and one that simply reads DANGER. Around the same time, craniometry was being performed on the skull of Native Americans. Some measurements found that their brain volumes were smaller than Europeans, suggesting they were slower to learn. This "discovery" justified the annexation of their land into the expanding new country and the consequent forced displacement out of their homes.

Today, most people will not publicly admit to holding any kinds of racial biases, especially not while under the scrutiny of scientists. To reveal these unconscious beliefs, researchers developed the Implicit Association Test (IAT), a test that can be easily conducted using some computer software and a keyboard. First, an image will appear (let's say it's a dog). The participant must sort the image into one of two categories. Push Q if that image belongs to the category on the left (mammal) but push P for the category on the right (reptile). After the response has been made, new images appear sequentially (snake, cat, mouse, lizard) while the categories remain on-screen while the person sorts the parade of animals (Q, P, Q, Q, P). In the next trial, the same images are shown in a randomized order, but the categories have changed. Instead of some phylogenetic sorting game, the categories are value judgments instead (good = Q vs. bad = P). The third task is the congruent trial, where two descriptors show up on each side. Now the participant is expected to push Q if the image belongs to one of two categories (a dog may be either mammal or good). This trial typically doesn't use much thought and therefore has quicker response times since we associate dogs with both mammal and good. Fourth is the incongruent trial, the critical trial that reveals our unconscious biases. Here, the person pushes Q even though the value descriptors are switched. (Is a dog a mammal? Click Q even though the word "bad" appears under on the left.) This task is difficult, usually causing the respondent to slow down, and these changes in response times might betray an unconscious bias.

The stimuli and the categories can be modified to test for our several different biases, demonstrating the powerful flexibility of the paradigm. We might use the IAT to see if people associate animals with other abstract behavioral traits. Do you believe dogs are loyal but snakes are treacherous? You might respond more quickly on a congruent trial. Show the name of a political philosophy. Do we have a tough time pairing them with justice but it comes easy to associate them with authoritarianism? Do you think people of certain skin tones are inherently better suited for certain activities? In the congruent trial, you might quickly push Q if it is the face of a Black person if it is paired with the word "athletic." A different congruent trial might pair the face of an older person with the word "slow." Another face: do we quickly associate men with "boss"? That same implicit bias might lead to a slow response when pairing a female name with words used in science, technology, engineering, and mathematics professions. Now here's an image of a same-sex couple. Is a heterosexual person quicker to associate them with the word "horrible"? Here's the name of a religion: do we associate it with acts of righteousness or terrorism? Are skinny, beautiful, lighter-skin-toned people better at their jobs?

As you can imagine, with a little creativity, the IAT can be modified to pull out some really icky "-isms" and "-phobias" we didn't know we had.

When performing the IAT of common stereotypes, such as those of race ("Is the name Tyrell black and unpleasant?") or gender ("Is Mary a woman and emotional?"), all sorts of hemodynamic changes pop up in fMRI imaging, with hot spots centered around the prefrontal cortex. Considering that these prefrontal cortex circuits function as the brakes that rein in the more primitive desires (see the previous chapter), it makes sense that they are involved in these tasks. If we could hear what these circuits were thinking, the dialogue might sound like this: "The primitive brain wants to conserve energy by making the quickest judgment possible. But do I really want to look like a racist or sexist in front of all these scientists? Let me make sure to suppress those socially distasteful responses and select the option that aligns with a less judgmental view of people." This processing takes time, which translates into the delay when the task is being performed.

On a separate race-based task, we tend to empathize more easily with people who look like us. Two populations of participants were selected. They identified as either White Caucasian (Italian) or Black African (born in African countries but having lived in Italy for an average of seven years). During fMRI scanning, they were then shown videos of hands of different skin colors, either that of their own race, the other race, or a digitally altered control condition of an unnatural tint of purple. Then the hand was either gently prodded with a harmless cotton swab or lancinated with a hypodermic needle. Some activity patterns showed the totally expected and unsurprising signals along the pain matrix. Seeing any hand in pain, independent of perceived similarity to that hand, produces a robust signal difference in the postcentral gyrus, which contains the somatosensory circuits. Another region with a strong signal is the premotor cortex, involved in motor planning, as if to communicate "don't just sit there you fool, pull your hand away before you get stabbed!" But the most interesting findings were made when comparing between the same-race and the other-race conditions. Two major loci of activity stood out.

First was an area in the right inferior temporal cortex. These areas are part of the ventral stream of visual processing, one of the two output pathways through the cortex that help us interpret what we see. Contained within this lobe are circuits for recognizing faces, places, and objects. Most relevant in this experiment is the extrastriate body area, which informs us about body ownership. Demonstration of the role of extrastriate body area can be done using the rubber hand illusion, practically a staple of every college psychology curriculum. Here, a person watches a detached rubber hand on a table being gently stroked by a paintbrush while their real hand, hidden from sight, is also stroked gently. After a few moments, the person starts to believe that the fake hand is theirs. Once this transference happens, the fMRI signal at the extrastriate body area increases.* This fMRI signal is greater in the same-race condition, suggesting that the viewer embodies the hand in the video.

The other area of interest was in the anterior insular cortex, which mediates incoming tactile signals and initiates an appropriate emotional response. This process, when applied to others, is empathy. Differences in insula signaling are seen in psychopaths, who are diagnosed in part by their low-empathy and highly callous, unemotional behaviors. In the same-race condition, the insula signal increases, but in the other-race condition, that signal decreases.

One possible interpretation of the results from the imaging study suggests that we imagine other's pain like it is our own—but only if they look like us. We may be evolutionarily wired to empa-

* This is also around the time that the cruel psychology professor pulls out a hammer and smashes the rubber hand, often resulting in a profound sympathetic nervous system response and an accompanying yelp of non-pain.

thize with those who are like us. It's easier to picture our own hand being swabbed or stabbed, and we are more likely to imagine the pain happening to us. But those same circuits intentionally dampen when we don't relate to the person. In doing so, we are spared the burdensome cognitive load of becoming invested in the plight of Them. Brain circuitry is difficult to change. This empathic dampening may have played a role in authors in the 1800s who suggested that pain tolerance is higher among African peoples as an excuse to justify their enslavement and cruel mistreatment. These beliefs have unfortunately been carried forward to the modern day, unknowingly fueling a set of commonly held myths about pain in Black patients in the clinic. One study revealed the pervasiveness of these falsehoods even among the highly educated: 8 percent of first-year medical students believe that the nerve endings of Black patients are less sensitive, and up to 40 percent of them believe that Black patients have thicker skin. These perceptions stay with some of these physicians when they enter practice. Black patients are more often undertreated with analgesics, and when they are given medication, they are given smaller doses.

These are just the beginning of the many health disparities between minoritized people and others.

> The Harkonnens sneered at the Fremen, hunted them for sport, never even bothered trying to count them. We know the Harkonnen policy with planetary populations—spend as little as possible to maintain them. —Frank Herbert, *Dune*

Inspired in part by nomadic hunter-gatherer peoples native to the Kalahari Desert, the Fremen are the natives of Arrakis. Although human, Fremen have adapted to a harsh environment that many readers find inhospitable. Water is their most valuable resource. They wear stillsuits, which allow them to reclaim the body's water lost through bodily functions, such as urinating, sweating, and even breathing. A diplomatic meeting nearly ends in bloodshed after a Fremen leader spits on the table, an act of respect in their culture. They have mastered their domain, but they are still looked down on as being savages, never earning the respect of the Harkonnen colonizers who once controlled Arrakis.

Belonging to a minoritized population, whether on Arrakis or on Earth, has effects on a person's health.* The difference in life expectancy in the United States when stratifying by race is more than a decade, with non-Hispanic Asians on the longest end and Native Americans or Alaskan Natives on the shortest. Obesity, a risk factor for cardiovascular illness and diabetes, is higher among Black and Hispanic Americans compared to White Americans. Black Americans are more likely to develop Alzheimer's disease or a related dementia, and when affected, they have more severe cognitive impairment and a more profound loss of autonomy. Some minoritized populations have worse outcomes after traumatic brain injury, are at a higher risk of psychosis, show increased markers for cerebrovascular diseases, and have significant thinning of cortical and subcortical gray matter as well as shrinkage of white matter tracts.

These health disparities are not usually driven exclusively by the "genes of race." Yes, there are some health conditions that are inherited and tend to run alongside certain populations. For these inherited diseases, changing environmental exposures or risk factors goes only so far toward helping patients. For most health conditions, however, the role of genetics is likely only one of many factors that explain any prevalence differences between the healthy and ill. For example, consider the link between Alzheimer's disease and the APOE4 gene. The APOE4 allele is the strongest risk factor for the onset of Alzheimer's disease, that is, at least in

* Even after controlling for sandworm-related deaths, which surprisingly doesn't have an ICD10 code yet.

Americans: the allele is prevalent in Nigeria, but the disease is rare. Considering the role of its protein product apolipoprotein-E in processing cholesterol, it is possible that this specific gene isoform, combined with a Western diet high in fat, increases Alzheimer's risk, while the Nigerian diet, traditionally high in fiber, negates that risk. With respect to the heightened likelihood of psychosis, this is present only when the patient is part of a socially isolated minority. Living in a community with a higher density of other people of the same ethnoracial background mitigates that risk, possibly through having a stronger, more understanding support network that can better connect with those from a shared cultural background.

And if pseudoscientific principles of "genetic superiority" and the effects of being minoritized aren't harmful enough, there is a suggestion that scientists aren't doing enough to combat racial disparities in research. In fact, many brain studies by their very design have some degree of built-in racial bias. One of the common diagnostic brain-imaging tools, the EEG, gets the strongest and most accurate readings of neuronal activity when the electrodes of the device are placed as close to the brain as possible. Therefore, the electrode pads need to go right up against the scalp to give the best adherence to the skin. The cleanest recordings are conducted in people who are bald. The most difficult signals to record are from people whose hair is curlier, more tightly coiled, and less absorbent of the saline solution used as an electrolytic conductor, three characteristics typical among those with African and Caribbean heritage.

Another real-time imaging technique, functional near-infrared spectroscopy (fNIRS), also requires close contact between the device and the scalp, once again introducing the same challenge when sampling from people with curly hair. An added complication with this technique is skin complexion. As the name implies, fNIRS relies on the emission of light waves. The light waves are scattered differently by oxygenated versus deoxygenated hemoglobin, which can then be detected as a proxy measure for regional changes in brain activity. Generally, infrared wavelengths are better than other spectra at penetrating biological tissue. But these waves still get attenuated by the higher levels of melanin in pigmented skin, producing unreliable readings in South Asian, Southeast Asian, and—yet again—Black patients.

Shortcomings in imaging technology aside, research participation has a representation mismatch issue that introduces difficulties with generalization. The UK Biobank contains extensive medical information from more than half a million participants recruited between 2006 and 2010. All these people are selfless donors who volunteer their time to science. They answer pages of questionnaires and sit through extensive interviews with experts. Some of them get diagnostic testing, such as MRI scans, heart ultrasounds, and whole-body bone and joint scans. Others contribute all variety of bodily fluids to the bank, which are stored in freezers equipped with a robotic arm that can pull a test tube from the shelf among 20 million other tubes. The end goal of all this work is to collect information about human genetics, lifestyle factors (occupational exposures, health history, substance use habits, and so on), and physical characteristics (grip strength, breath strength, bone density, and so on) to uncover correlations between those 10,000 or so variables and health outcomes for diseases of middle and late age. According to predictions, in the 20 years following the initial collection of these data, the research team expects to see 30,000 cases of Alzheimer's disease, 20,000 strokes, and 14,000 cases of Parkinson's disease. Best of all, these data are made publicly accessible for researchers around the world to use. More than 9,000 academic publications have used data from the Biobank. Among such data are descriptions of blood biomarkers that predict various forms of dementia, novel gene sequences associated with depression, and risk factors that predict substance use disorders. With a data set so massive and the capacity to track outcomes over decades, it is the most important prospective health database in the world.

In the White Western world, that is. The Biobank collected samples from across 22 sites, representing a mix of both rural and urban populations. But the sample is 95 percent White participants, a demographic group that is a global minority making up only 12 percent of the population. This is just one of many cases where the sample contains a narrow slice that is misrepresentative of global society. Of the psychology studies published in top-tier behavioral science journals in the early 2000s, up to of 96 percent of participants were from the United States, Europe, Australia, and Israel. A further stratification reveals that nearly 68 percent of participants were from the United States. In fact, this discrepancy between these participants and the rest of the world is so common in neuroscience and psychology research that there's even an acronym for these overrepresented people. There's a strong chance that a participant from any given brain research study is weird—well, not just weird but rather WEIRD. That is, they were likely raised in a society that is Western, Educated, Industrialized, Rich, and Democratic. In other words, discoveries made in the typical research lab don't represent any absolute truths about the nature of *Homo sapiens*. Instead, they likely reflect the way that WEIRD people have been acculturated to think.

Certain behavioral tasks are wonderful for extracting culturally dependent ways of thinking, especially for societal-level beliefs regarding interpersonal relationships. One such example is the Public Goods Game. Here, each of four players is given an amount of money. They have the option to contribute that money to a public pool where it is multiplied by some factor and then redistributed to all players. To maximize total profit, all players should contribute their entire allotment. Recognizing this, one player may act selfishly and contribute zero to the pool, thus reaping pure profit while risking nothing. Most participants, when given the power to punish other players in future games, are willing to pay a small amount of money to punish those good-for-nothing freeloaders, loafers, and parasites who are a drain on the system. Hopefully, by punishing them once, it will prevent them from trying to take advantage of others' generosity in the future. This thought process appeals to a cross-cultural ideology of justice and appears in all 16 of the cultures assessed, including people from a diverse selection of cities such as Boston, Seoul, Copenhagen, and Istanbul. But there is one behavior that WEIRD people find difficult to rationalize. Some participants, specifically those from Muscat, Oman, Athens, Greece, and Riyadh, Saudi Arabia, chose to punish the most generous players for contributing to the pool even at a personal loss.

The way objects get classified differs between cultures along the global West/East division. Americans of European heritage judged groupings of abstract shapes based on a rule they invented, such as whether the shape had a swirly string on top or pointy edges, regardless of the other images surrounding the shape in question. East Asians, all international students* for whom Chinese or Korean was their native language and primary cultural exposure for the majority of their lives, categorized these stimuli based on how similar they looked as a whole to the group. The third group, Asian Americans, who may have grown up with Eastern cultural elements but were socialized by Western influences, showed no strong preference toward rule or group similarity, essentially bridging the cultural gap.

WEIRDness doesn't apply only to the domains of interpersonal or inter-object relationships. It can even go all the way down to the most basic levels of some of our elementary functions, such as vision. The Müller-Lyer effect is an illusion where two lines of the same length appear to have different lengths depending on the direction of the tails that extend off the end of the lines. A cross-cultural study demonstrated that this misperception is most pronounced among WEIRD people. One population of participants, people

* All undergraduates, again demonstrating a bias in research toward populations that fit the criteria of E, I, and R.

from Evanston, Illinois,* didn't perceive the two lines as being of equal length until there was a 20 percent actual difference in length. On the other hand, people from hunter-gatherer or subsistence farmer societies are practically immune to the illusion.

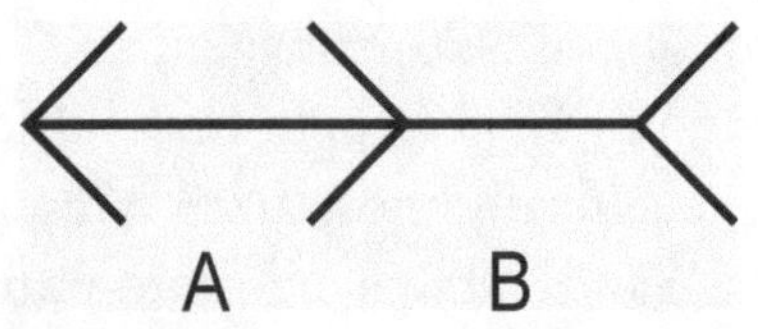

Figure 10.1. A Müller-Lyer stimulus. If you think the horizontal segments A and B are the same length, congratulations, you're WEIRD. Segment A is around 50 percent longer than segment B. Welcome to the club. CREATED BY AUTHOR

The thing about WEIRDness is that it is not a consequence of aspects encoded in genetics. Rather, it is a product of institutions and culture, both of which are shaped by the values of being WEIRD. There's not a sequence in our WEIRD genes that codes for a deficit in the ability to accurately judge the lengths of lines. WEIRD people fall susceptible to this illusion because of our heavily industrialized infrastructures and an architectural obsession with neat little right angles.

Sorry, neither. —Lieutenant Uhura, *Star Trek: The Original Series*†

In 1965, a confident, competent, intelligent Black woman was seated at the console on the USS *Enterprise*. Her name was Lieutenant Uhura. Her role as Starfleet's communications officer seldom gave her the chance to play the heroine, but she was still a

* As a former Evanston undergrad, I can confirm that the student body there is both WEIRD and weird.

† Season 1, episode 4. A sweaty Captain Sulu gets space drunk, loses his shirt, and tries to save Uhura from a hallucinated threat, calling her a "fair maiden" in the process. Her two-word rebuttal was a victory for both people of color and women. Legend has it the line was ad-libbed.

capable Black Woman in Space in an era where the role of actresses who looked like Nichelle Nichols was relegated to being the help.

Nichols was a born singer and stage performer. For her, *Star Trek* was intended to be only a stepping stone toward something else entirely. Near the end of the first season, she was given her dream opportunity: a role in a show headed to Broadway. With this news, she handed her resignation letter to producer Gene Roddenberry, who insisted that she at least take the weekend to reconsider. Days later, she appeared as a celebrity guest at an NAACP fundraising event. One of the organizers informed her that someone who claimed to be her biggest fan was desperate to meet with her. In her mind, she imagined some young teenage Trekkie who was likely to greet her not with a standard "Hello" but with an emotionless "Live long and prosper" accompanied by the Vulcan salute.

To her surprise, that fan was Martin Luther King Jr.

During their conversation, Nichols casually mentioned that she would miss her costars, confiding in him that she planned to quit *Star Trek*, to which he sternly replied, "You cannot." *Star Trek* was the only show he and his wife allowed their kids to stay up past their bedtimes to watch, and it was because of Nichols. In King's view, Nichols's portrayal of Uhura was too important for the civil rights movement for her to walk away. She was a visible role model, one who embodied a future that embraced equality. "For the first time," he said, "the world is seeing us as we should be seen."

King was probably right. After this conversation, she would return to Roddenberry with the news he wanted to hear, as his vision of the future was one that showcased diversity. She would continue as the recurring polyglot comms officer for the duration of its three-season run. She will inspire many other Black women who will now see themselves in a position typically given to White men. A 10-year-old Whoopi Goldberg will exclaim to her mother, "There's a Black lady on TV, and she ain't no maid!" Years later, Goldberg will play Guinan on several episodes of *Star Trek: The*

Next Generation. In 1977, Nichols will appear in a NASA recruitment video where she will be given a tour of the training facility and the inside of the space shuttle cockpit. "This is your NASA," she will say, inspiring both Guion Bluford and Sally Ride to join NASA in the following year and Mae Jemison* a few years later. All these people will break racial and gender barriers by breaking free from Earth's gravity.

Nichelle's story serves as a reminder of the value of representation and the importance of exposure for changing stereotypes. Broadly speaking, this idea refers to the contact hypothesis, the observation that prolonged, positive interactions with people of different backgrounds decreases harmful prejudice against others of those backgrounds. There are caveats, however. There must be equality in status, they must work together to achieve a common goal, and there should be support for the forming of bonds from the related authority infrastructure. For example, the U.S. Army during World War II was officially segregated. However, when conditions required the integration across racial lines, White soldiers had greater positive attitudes toward Black Americans.

Some experimental research suggests that subconscious biases, such as the own-race effect, can be overcome by exposure, suggesting that facial recognition bias is a product more of acculturation than of biological hardwiring. This was demonstrated by recruiting children from multiracial societies. For example, one study demonstrated that French natives are better at recognizing Anglo faces and that Korean natives are better at recognizing East Asian faces, perfectly in line with what the ORE predicts. But Korean children adopted by French families surrounded by French people daily scored better at recognizing Anglo faces compared to East Asian faces. Similar effects have been shown

* Sometimes, everything comes full circle. After returning from space, Jemison will guest star on an episode of *Star Trek: The Next Generation*.

for Caucasian Brazilians, who have a greater difficulty recognizing Japanese faces, while Japanese Brazilians do not. The same holds for African students in South Africa, who were better at recognizing White faces than their African nonstudent counterparts, a disparity attributed to their frequent exposure to White students and people in positions of authority or power at the mostly White University of Cape Town.

Despite all the complexities of human societies, contact hypothesis is surprisingly backed by some nonhuman evidence as well. Under the right circumstances, even members of an out-group may be treated like an in-group, suggesting that antisocial behaviors are not inherently baked into genetic programming. A clever rat model for empathy was developed by Peggy Mason* and her graduate student at the time, Inbal Bartal. Here, a rat is restrained uncomfortably within a small clear tube that can be opened only from the outside. Another rat, free to roam about the larger cage, is put outside the restraining chamber. Through random trial and error, the free rat learns to flip open the door for their fellow trapped rat. Once the animals figure out how to do this, releasing the friend becomes the first thing they do on future trials. Then they tested to see how contact with other types of rats changes this behavior, using different strains of lab rats with minor differences: Sprague-Dawley rats are albino with white fur and bright red eyes, while Long-Evans rats, or black-hooded rats, have black fur all around their head and face and brown eyes. When put into this behavioral test, a Sprague-Dawley would liberate a Long-Evans if they were cage mates but not if the Long-Evans was a stranger. Furthermore, the Sprague-Dawleys even generalize their willingness to free other novel Long-Evans strangers but only if his cage mate is also a Long-Evans.

* I don't name-drop very often, but Mason was my systems neuroscience professor in grad school.

Mice have demonstrated an empathic-like response that depends on familiarity, suggesting that they too are aware of in-group and out-group dynamics. Humans exhibit emotional contagion when we unconsciously mimic the behavior of others to reflect a similar cognitive state—think of how people squirm in their seats during gratuitous depictions of violence in a gory slasher film. Pain is also emotionally contagious in mice, particularly through observations of writhing, the bodily movements in response to a mildly noxious irritant. To study this mimicry, mice were placed into a clear tube for observation and then injected with acetic acid. In response, they would twist their body, stretch their belly, and kick their hind legs. They were tested in pairs where both mice would be injected but only one would receive the acetic acid. The saline-injected mouse would writhe in empathy if the pair were cage mates, the mouse equivalent of "I feel your pain mouse-brother. We'll get through this together."* When a stranger, one who had little to no shared experiences, was put into the test, there was no emotional contagion. The implications here are twofold. First, the ability to convey sympathy toward someone in a state of pain depends on exposure. Second, the prosocial benefits of empathy probably came up through evolution before humans and other primates.

And in further support of how evolution has maintained the ability to rewrite empathy toward out-groups in light of exposure, consider weaver ants (*Oecophylla smaragdina*). Like other eusocial animals, they live in colonies, sometimes with up to half a million other ants. They launch precision rescues of fellow ants caught in life-threatening predicaments. But at the same time, they are highly territorial, enforcing their domain from rival weaver ants by constructing smaller defender colonies on the outskirts of their

* Although the total lack of empathy is unsettling, overly empathic displays of emotion can be equally disturbing. Think of the scene in *Midsommar* when the women of the village were crying feverishly alongside the protagonist.

boundaries. Weaver ants across the Solomon Islands just off Australia were followed closely to see what they would do when they encountered a fellow ant trapped in a spiderweb. Sometimes, they rescue the helpless ant, chewing at the sticky silk snares until the ant is free. Other times, they will see an enemy in peril and actively attack them, biting at the helpless victim until they die. They always tried to rescue an ant from the same colony. Given their territorial nature, it was surprising that these ants rescued ants from other colonies around 40 percent of the time. Their likelihood to rescue outsider ants depended on how far the outsider's colony was from the rescuer's. Some degree of familiarity, provided by scant exposure to pheromones, encourages cooperation in a time of crisis.

Optimistically, these results imply that frequent interaction with people from different backgrounds enhances our empathy toward them, narrowing that distance between Us and Them. Of course, these nonhuman models are not truly analogous to the human condition, nor do they consider the multitude of complex variables that shape our world. Rat models don't worry about having their fur color artificially lightened on magazine covers, and mouse casting directors don't get criticized for selecting darker-skin-toned actors to play guards and housekeepers to appease the sense of Eurocentric beauty norms.* But they do suggest one way by which negative stereotypes can be broken.

---

* *Vogue* magazine and *Crazy Rich Asians*, respectively.

CHAPTER ELEVEN

DIAL 56 TO SMELL BURNT TOAST

What Brain Stimulation Tells Us about Free Will

At any given moment, there is an expansive set of any actions available to you from which you may choose. In other words, we have free will. Some philosophers subscribe to determinism, the idea that free will is an illusion. According to this train of thought, every action can be predicted. Perhaps the actions have already been preordained by the omniscient will of a powerful deity or guided by the movement of the celestial bodies. Maybe the action is a result of our genes, our past experiences, and the specifics of the environment in which that decision is being made. To an outsider, actions may appear capricious, but determinists suspect that patterns of predictability emerge among the chaos.

Neuroscience entered the free will discourse in 1983. An EEG was placed on the heads of the participants. They were told that at any moment, completely of their own doing, they could move their finger. Surprisingly, there was an increase in neural signals up to a second in advance of the muscle movement. This signature change in brain activity came to be called the readiness potential since it was interpreted as the brain readying itself for sending the motor command signal down to the muscle. While these results did not definitively destroy free will, it certainly felt a lot like some unexplored aspects of determinism. Somewhere in the brain, proteins did something, leading to diffusion of molecules across a synapse, and a second later, twitch, finger movement.

More recently, Stanford professor and neuroendocrine expert Robert Sapolsky weighed in on the issue with his treatise on free will, *Determined* (2023). Suggesting that however we conceptualize free will, whatever we think it means, it's likely just a complex interaction of factors that guide our actions. Consider this analogy: if you know the exact edge curvature of a six-sided die, the coefficient of friction of the surface onto which it is thrown, the velocity and angle of release, the density of the air, and hundreds of other variables that can be accurately calculated using the immutable laws of physics, then, technically, you can predict the outcome of every single roll of craps. In humans, it's a bit more complicated, but the same premise applies. Knowing precisely what genes you were born with, the role of the trillions of environmental influences that shaped your molecular milieu during development, and every decision you've ever made at any moment in the past allows your next action to be predicted. This depiction of free will should feel wildly theoretical. We're not at the point where biology is granular enough to predict human behavior.

Determinism has a certain draw to it. It lets us be guilt free, as in "your actions are not your own." You're always acting under duress, a hostage to your neural circuitry and the genes that are turned on or off at any given moment. This obviously has some weird implications for the judicial system. Neither "the stars made me do it" nor "my brain made me do it" would ever convince a jury of a person's innocence.

But at the same time, no one wants to feel like the puppet of some ventriloquist. The denial of free will has a certain horror to it. And there are few things scarier than the mind manipulation tricks developed by evolution.

> I knew what to do, what to expect. I had been told all my life. I felt the familiar sting, narcotic, mildly pleasant. Then the blind probing of her ovipositor. The puncture was painless, easy. So easy going in. She undulated slowly against me, her muscles forcing the egg from her body into mine. —Octavia Butler, *Bloodchild**

Like many experimental biologists and horror aficionados, I pride myself on having a pretty strong stomach when it comes to dissection, surgery, and graphic depictions of gore in movies. But there is something so absolutely repellant about parasites that just thinking about them makes me wish I hadn't eaten.

Author Octavia Butler had a similar distaste for this corner of nature. After scoring a book deal for her *Xenogenesis* trilogy, she traveled to the Peruvian Amazon in search of further inspiration. Before her trip, she was warned about the botfly. Physically, the botfly has the same cute appeal as a chubby bumblebee. It's rotund and a little bit fuzzy and has disproportionately huge black eyes. But don't let its appearance distract you from its terrifying reproductive behavior. It lays its eggs, maybe directly on the hosts, which include cats, dogs, deer, horses, monkeys, and people. These eggs then live underneath the skin, hatching into larvae that feed on the flesh of whatever mammal becomes infested. If left untreated, the larvae grow over the course of months before erupting through the skin, sometimes spilling out by the handful. For most people, the thought of being a host for a botfly infestation fuels nightmares. Butler turned that fuel into the Nebula and Hugo Award–winning *Bloodchild*.

Still haven't lost your lunch? Let's see if I can share more about other parasites that will change that.

---

* In *Bloodchild* (1984), humans live under the protection of a sentient alien species called the Tlic. A young boy is chosen to be the bearer of the eggs of a Tlic, a great honor despite the risks involved in birth. The Tlic are both politically and physically dominant, and Butler emphasizes this power dynamic while exploring the phobia of male impregnation and critiquing societal expectations about women's role in society. It might be the first piece of feminist literature to feature centipedes.

If *Arachnophobia* inspired a deeply ingrained fear of spiders that won't go away even knowing that more than 99 percent of spider species are incapable of causing severe physical harm to humans, the following animal relationship may help you conquer that fear. In this story, the spiders aren't the bad guys. If you allow yourself to, you may find yourself sympathizing with the Eight-Legged Freaks you once feared. Granted, by the end of this vignette, you'll likely have traded one fear for another.

Native to Costa Rica is an orb spider called *Plesiometa argyra*. These spiders spin the webs you imagine in your head when you think of spiderwebs: a circular central hub where the spider spends most of its day, a handful of outwardly radiating spokes connecting to nearby leaves and stems, and several densely packed concentric rings made of sticky silk for catching small insects that are unfortunate enough to fly into their home. Also found in this ecosystem is a horrific little wasp called *Hymenoepimecis argyraphaga*. Like other wasps, it has a slender, needle-like body and translucent wings. At its rear is a long, multi-segmented tail that looks like nature was playing around with Legos and mixed up the wasp and scorpion boxes.

As with the botfly, this wasp's reproductive cycle is the most distressing thing about them. When two wasps love each other very much, the female looks for a place to lay her eggs. Instead of choosing somewhere unobtrusive, such as an underground nest or a tubular cell made of mud as other wasp species do, this wasp finds an orb-weaver spider, holds it down, stings it repeatedly until it becomes paralyzed, and then deposits her eggs. Then, after the paralysis wears off, the spider goes about its normal life for a few weeks. But something changes in the brain of the spider. Instead of building a normal, bug-catching web, some chemical released during the process of the parasitic infection causes a change in its web-spinning pattern, where they now build a reinforced dragline between a few fixed points. This web covers less total area, but it is structurally able to support more weight right at the center. This type of web is called a cocoon web because of what happens the following night. The wasp larva matures, emerges from the egg, sucks the spider dry of nutrients,* and then forms a cocoon that dangles from the center of the web.

Good-bye arachnophobia, hello spheksophobia.

Several other species of wasp adopt similarly nefarious child care strategies. In Japan, the *Zatypota maculata* wasp launches a similar web-hijacking assault on the *Nihonhimea japonica*, as does the *Reclinervellus nielseni* and its spider host *Cyclosa argenteoalba*. In Ecuador, *Zatypota* wasps exert a similar mind manipulation on the *Anolesimus eximius* spider, a social spider. After the egg has been laid, these spiders leave their homes and start spinning cocoon webs, which the emerging larvae will use to molt after they, of course, drain the spider dry. The *Dinocampus coccinellae* wasp targets ladybugs. Within weeks, the larva erupts from the belly, then quickly builds its delicate cocoon under the protective red and black carapace of the paralyzed ladybug.

Onto another stomach-churning battle of nature.

This creature is *Leucochloridium paradoxum*, commonly called the green-banded broodsac. It belongs to the phylogenetic grouping of flatworms, nasty egg-laying parasites that can infect all sorts of animals, including household pets and their owners, sometimes capable of growing up to 30 feet (nine meters; also, YUCK) inside your intestines. This specific flatworm doesn't care too much about people. Instead, their preferred host are birds, where they spend much of their mature life feeding on the stuff that passes through the bird rectum. When they release their eggs, those eggs end up in the feces that falls to the ground. Here, the eggs may take up residence in a secondary host, such as amber snails, that feed on the feces.

---

* If you know Latin and paid close attention to the scientific names of these two species, you might have picked up on the spoiler to this horrific tale. The spider, *argyra*, is eaten by the larvae of the wasp, *argyraphaga*.

This is where the tone shifts from just gross to sinister *and* gross. Once inside the snails, the eggs hatch and turn into sporocysts that begin to take over the eye stalks of the snail. Normally, amber snails have thin brown eye stalks that they can use to retract their little bulging eyes into the body for protection. When the broodsac takes over, it climbs into the eye stalks, engorging them. The sporocysts are brightly colored, usually with multiple stripes of green, yellow, and white. Under the influence of the flatworm, the once-shy snails now climb to high surfaces atop leaves and branches. The sporocysts start undulating, creating a pattern of flashing colors evocative of a juicy caterpillar out in the open. The eye stalk is practically a giant neon "Diner Open" sign for birds flying overhead, which happily swoop up their prey to continue the twisted cycle. This behavior gives *L. paradoxum* the ominous moniker: snail-manipulating flatworm.

More famous than these other terrifying parasites is the protozoan *Toxoplasma gondii*. Wasps and flatworms are bigger organisms and therefore have specific demands on an ideal ecological niche, so factors such as climate and resource availability limit where they can thrive. But *T. gondii* are microscopic and survive inside gastrointestinal cells of practically every avian and mammalian species. Because of this, it is unsurprising that infections are found worldwide, from the Amazon to the Arctic. Thankfully for us, humans are not their preferred hosts. That dishonor belongs to cats. It is in the cat digestive tract where the cysts fully mature. The parasite is then shed in feces, which can leak into the water supply and contaminate food sources, later to be taken up by the many other intermediate hosts in the ecosystem, such as cattle and swine.*

As far as studies on these intermediate hosts, the influence of *Toxoplasma* infection on mouse behavior is the best studied. Here

* House cats are at an increased risk of picking up *Toxo* from the soil or their prey when they play outside, another excellent reason to keep these furry little killing machines indoors.

is where the parasite shows its psychic mind control ability. Normal, uninfected healthy mice have an inherent, healthy fear of cats, even among mice that have been born and bred in laboratories far removed from nature. This can be demonstrated in a test environment where one side of the box contains a dish of bobcat urine. In the control condition, there is a dish of rabbit urine, a novel aroma but that of a non-predatory animal. Normal mice cower from catty aromas, which is why homeowners dealing with mouse intruders scour the aisles at the local hardware store for bobcat pee. But give these mice a *Toxoplasma* infection, even of the mildest variant, and they lose their fear of that pungent aroma. This behavioral modification is permanent, as the behavior persists even after the parasite has been cleared from the mouse and brain inflammation returns to baseline levels.

The ecological implications behind this observation are nothing short of science fiction. Somehow, *Toxoplasma* convinces the mouse into believing that cats aren't the deadly apex predators their sharp claws and lightning reflexes make them. Mice become less fearful of potentially threatening environments, leading them to run out into the open, where they become easy prey for cats. After that little morsel, *Toxo* takes up residence in the cat intestinal tract once again, and the cycle begins anew. The healthy rodent becomes brainwashed by the parasite, increasing risk-taking behaviors—sort of a self-destructive *Mousechurian Candidate*.

Given what is already known about the neural circuitry involved in fear-related processes, it is unsurprising that one mechanism behind this behavior lies in the basolateral amygdala. A *Toxo* infection leads to lower blood levels of circulating corticosterone, the main neurohormone that is involved in the stress response. A chilled-out mouse is unafraid and will likely disregard warning signals in their environment. *Toxo*-infected mice also have shorter dendrites of the principal neurons of their basolateral amygdala, which aligns with decreased signaling through these circuits.

As with mice, *Toxoplasma* can live in humans as an intermediate host. Most of the time, we pick it up from drinking contaminated water or eating undercooked meats and raw vegetables. Alternatively, it can be passed during birth from an infected mother to child and occasionally after an organ transplant if the donor was infected. These routes of infection have led to high prevalence rates, with estimates of people in the United States carrying some *Toxoplasma* load around 20 percent. Globally, the numbers are higher with hot spots in Latin America and Eastern and Central Europe. In some tropical areas where the parasite thrives, prevalence estimates reach up to 75 percent. Thankfully, most people with the infection are asymptomatic since a healthy immune system can suppress the negative effects of the parasite. But for the immunocompromised and newborns, an infection can lead to vision loss, acute illness, fever, seizures, and even respiratory and heart failure.*

While these effects on the organs are dangerous and sometimes even fatal, it is infection of the central nervous system that makes *Toxoplasma* especially insidious. The risk of developing various mental illnesses is higher among those harboring the parasite. For example, a person with a *Toxo* infection is almost twice as likely to develop schizophrenia than those without. Antipsychotic medications such as haloperidol have been thought to ease the symptoms by blocking the dopamine D2 receptor. But these same drugs are toxic to the parasite, suggesting they work by clearing the body of *Toxo* rather than by changing neurotransmitter signaling. Among patients with schizophrenia or OCD, nearly twice as many of them are positive for *Toxoplasma*, suggesting this parasite is a risk factor in the disease. Also, suicides and suicidal ideation is increased among people with *Toxoplasma* infection by approximately 50 percent.

---

* Pregnancy decreases immunocompetency, which is why health advice suggests that pregnant women delegate the chore of changing kitty litter to someone else.

A *Toxo* infection also changes people at the subclinical level. In men, the infection makes them more suspicious of others. In a laboratory setting, they are less likely to put their signature on a blank sheet of paper, allow others to hold their wallet, or drink a strange chemical given to them by the experimenters. Infected men also had less self-control as measured by how often they clean their household or separate their waste into paper and plastic bins, how many events they have planned over the next few days, and one sneaky little measurement of how late they arrived to the experiment. These same infected men also had clothes that were less tidy. For women, however, *Toxo* infection status did not explain any differences in self-control or appearance measures.

The mental image of the "crazy cat lady" is all wrong. The stereotype needs to be updated to "sloppy, paranoid cat men."

Toxo infection also has psychomotor consequences. In people being tested on a simple computerized reaction time test, where a change in an image on the computer screen requires an input on the keyboard, those with the infection had slower reaction times. Furthermore, the longer the person had the infection, the worse their reaction times were. Other tests have shown that *Toxo* increases impulsivity and aggression and decreases rule following. These psychomotor changes are all part of *Toxoplasma*'s insidious reproduction and survival plan. Less reactive mice means that cats have an easier time catching their prey, and increased impulsivity means that they are more likely to dart out into the open. Just imagine *The Happening* but with mice.

Think of an everyday task that requires good reaction time and rule following while also benefiting from low impulsivity and aggression. For one such task, the odds of doing badly at it are increased by about double if you have a *Toxo* infection.*

* It's driving. *Toxoplasma* is much more prevalent among people who have been in major car accidents.

> Viruses can make us ill, but fungi can alter our very minds. There's a fungus that infects insects. Gets inside an ant, for example, travels through its circulatory system to the ant's brain . . . thus bending the ant's mind to its will. The fungus starts to direct the ant's behavior, telling it where to go, what to do, like a puppeteer with a marionette. —*The Last of Us* (2023)

Without ever saying the word "zombie," George Romero's *Night of the Living Dead* popularized the tropes that make up the modern zombie myth: flesh-decayed reanimated corpses, limbs stiff with rigor mortis, tenaciously searching for human flesh. Zombie films throughout the decades have been inspired by the fears of the zeitgeist. In the 1960s, people were preoccupied with the space race, having witnessed many "firsts" earlier in the decade before the movie's release in 1968. A news report in the film speculates that the zombies are a result of exposure to cosmic radiation brought back from a probe that circled Venus. The survivors bicker among themselves, mirroring the generational divide between the counterculture movement of the idealistic youth and the traditional values of the older adults. Racial tensions also ran high as the lead actor was a Black American who, after surviving the zombie threat, is fatally shot and killed by a heavily armed human mob.* The sequel, *Dawn of the Dead*, hit theaters a decade later in 1978. This time around, Romero wanted to critique the consumerism that was being hailed as an American virtue. So he had the survivors barricade themselves in a shopping mall to highlight the parallels between us and the zombies, both beings driven by a mindless desire to get more. George R. R. Martin's *A Game of Thrones* (1996) and the HBO adaptation (2011) gave us zombies in spirit but not in name. Called wights, these are reanimated beings from the frigid

* Through sheer coincidence, Martin Luther King Jr. had just been assassinated right as Romero finished his final edits of the film.

lands north of The Wall, and they embody the creeping threat of anthropogenic climate change. In *28 Days Later* (2002), the cause of the zombie outbreak was a mysterious infection, likely inspired by the fears of bioterrorism and the malintent of rogue actors in the Middle East.

At some point, fatigue hit hard. Zombie franchises piled all around us like a horde of, well, you-know-whats. *The Walking Dead* had six spin-off series. Zombies burst out of their expected horror settings, shambling into other genres with the romantic comedy *Warm Bodies* (2013) and the musical *Poultrygeist: Night of the Chicken Dead* (2006). The *Resident Evil* video game franchise has averaged one release per year since 2000. The trope was getting played out, providing inspiration for satires that poked fun at the genre with wonderful titles such as *Shaun of the Dead* (2004) and the Cuban film *Juan of the Dead* (2010). It would take something really special to stand out in this saturated market. That's where *The Last of Us* (2023) makes its mark. Based on the multiple-award-winning 2013 PlayStation 3 game, *The Last of Us* did more than carry George Romero's torch. These zombies ran with it.*

The thread underlying all of these zombie films, whether scary or comedic, is a reasonable speculation: the loss of free will leads to the utter collapse of society.

The mind-hijacking agent in *The Last of Us* is a fungus. While most fungi feed on the dead, the actual fungus that inspired the franchise, *Ophiocordyceps unilateralis*, feeds on the living. The opening sequence of the HBO series showed us the workings of this mind control fungus, appropriately named the zombie-ant fungus. This parasitic organism is found across tropical rainforests as diverse as Brazil, Thailand, and Australia. Their hosts, various species of carpenter ants, typically live high up in the canopies. Occasionally,

* This wasn't the first depiction of athletically gifted zombies, with Max Brooks's *World War Z* (2006) and *Train to Busan* (2016) showing some of the fastest, most agile zombies. Even the opening scene of *Night of the Living Dead* (1968) shows a reasonably spritely zombie.

they are forced to the ground if the leaves do not form a continuous bridge between trees. Once here, spores on the forest floor stick to the ants. They penetrate the ant's exoskeleton and make their way into the ant's brain. The fungus releases all sorts of chemical signals that influence ant behavior, making them behave in really bizarre ways. First, they lose control over their limbs, often causing them to fall to the forest floor. Then they climb to the perfect height, clamp tightly onto the stem of a leaf with their mandibles, and die in place. If you've never felt empathy for an ant before, just watch a time-lapse video of the horrific scene that comes next. Over the course of weeks, the fungus grows until it bursts through the empty husk of the ant's body, sometimes sprouting a long, thin mushroom tendril out of the head. Then, once the mushrooms have matured, they scatter their spores into the wind, ready to infect the next ants that venture across this mind control minefield.

The human mind is also susceptible to fungus-mediated behavioral changes, a fact we've known for a long time. A topic of debate among ethnomycologists is just how long ago those effects were first documented. A prehistoric cave painting in Tassilli n'Ajjer in Algeria shows a vaguely humanoid being wearing a blank, expressionless mask, adorned in mushrooms. Supporters of the mushroom theory offer the suggestion that this is a shaman, engaging in some kind of divine ritual. The drawing may date back to well in the prehistoric era, close to 7000 BCE. Another ancient piece of cave art is a drawing from around 4000 BCE, called the Selva Pascuala mural in Spain. This one shows a bull and a row of 13 mushroom-like objects to the right. Species of the *Psilocybe* genus, or magic mushrooms, grow in this climate, and some prefer growing out of bull dung. Other evidence suggests the mural was not a place of residence but rather a place of some kind of special significance. Mind-altering *Psilocybe* mushrooms also played a role in pre-Columbian Mesoamerican societies. Their art features mushrooms prominently, including mushroom shaped ocarinas, stone carvings of people with mushroom caps for heads reminiscent of Toad from the *Mario* franchise, people dressed in ceremonial garb holding fistfuls of mushrooms, and even paintings of mushrooms alongside human sacrificial rituals. Of course, it's completely possible that people may have taken shrooms recreationally for the intoxication, but they were most likely used for their entheogenic properties—the sensation when the user develops a heightened spiritual connection with the higher beings that are a part of the culture.

While psilocybin is the most popular psychedelic of fungal origin, keep in mind that LSD was chemically derived from the ergot fungus.* Project MK Ultra tested if LSD had any potential as a mind control agent, a goal inspired by fears of Cold War escalation. In theory, by giving LSD to an enemy combatant, an interrogator could reprogram a person's inhibitions, causing them to voluntarily reveal military secrets. The CIA was well aware of the limitations of the project, even acknowledging that "truth serum," while certainly a punchy phrase picked up by the media, is a misnomer. But it didn't stop them from dosing incarcerated people, sex workers, and even their own employees, such as a civilian biochemist named Frank Olson, who died by suicide after an MK Ultra–related LSD dosing (although conspiracy theories suggest his fate was not of his own choosing).

> He beckoned me to the window. I got up and looked out, and He raised his hands, and seemed to call out without using any words. A dark mass spread over the grass, coming on like the shape of a flame of fire; and then He moved the mist to the right and left, and I could see that there were thousands of rats with their eyes blazing red—like His, only smaller. He held up his hand, and they all stopped. —Bram Stoker, *Dracula*

* Hoping to prompt flashbacks to chapter 4 here.

One of the more recent and certainly most revolutionary methods of rodent mind control is not supernatural at all. In fact, it's almost all natural—as in, it comes from nature. In the early 2000s, scientists described a protein called channelrhodopsin-2 (ChR2). It was isolated from a single-celled green alga, *Chlamydomonas reinhardtii*. Like other plants, *C. reinhardtii* gets much of its energy through photosynthesis. But for this soil dweller to maximize the power of the sun's rays, it must move toward that light. Researchers discovered that the ChR2 in *C. reinhardtii*'s primitive eyespot organ serves as the signal transducer that tells the animal which way to move. When light is flashed onto ChR2, the protein undergoes a tiny biomechanical change. In a matter of milliseconds after exposure to light, cells that express ChR2 change their electrical properties, causing the algae to move toward the light. Using the massive toolbox of genetic manipulation strategies that have been fine-tuned over the decades, these light-sensitive ion channels can be put into neurons with extraordinary spatial precision, ignoring certain populations of neurons within the same physical area of the brain. Once these cells start synthesizing and expressing ChR2, a tiny fiber-optic cable can be surgically threaded into the brain, activating subpopulations of the brain using flashes of light. This stimulation strategy is called optogenetics.

Now scientists have developed rodent mind control powers that approach the supernatural talents of Dracula. Consider the following behaviors that have been manipulated using optogenetic tools:

- A mouse explores a large arena, sniffing and rearing up on its hind legs, displaying totally expected mousey curiosity. Then the light turns on. Suddenly, he starts sprinting at full pace, furiously blazing a counterclockwise circle like Mousain Bolt. Here, the ChR2 has been put into only one motor cortex. Activating these cells sends a strong activity signal down the spinal cord and through the nerves powering the muscles on the opposite half of the body. This asymmetric muscle activity causes the mouse to run circles in his arena, like a rower with one arm much stronger than the other.

- In their home cages, mice are given unfettered access to food. They are removed and put into a new test chamber in which sits a pile of food, which they saunter past completely uninterested. Then the light turns on. No matter where they are or how sated they are, they move quickly toward the food. The aggression with which they eat is impressive, on the same scale as a mouse that has been starved for 24 hours. This ravenous appetite is a result of activating a portion of the hypothalamus, a key regulator of basic homeostatic functions. Specifically, when the agouti-related peptide-producing neurons of the arcuate nucleus are fired up, this artificial hunger overrides the normal satiety signal.
- A two-room box serves as a testing chamber. Each room has its own distinct set of environmental cues, such as unique colors and floor textures. The mouse roams freely between both rooms, finding both equally enjoyable to explore. The mouse enters one of the rooms. Then the light turns on. They quickly learn there is something about that room they enjoy and prefer, while the other room offers nothing to them. Here, a population of dopamine neurons in the VTA are activated by the optogenetic manipulation, which increases their wanting while in that chamber.
- Picture a peacefully slumbering mouse. Although they sleep in bursts rather than in long stretches like we do, they spend a total of half their day dozing. Then the light turns on. He suddenly jolts awake—not in pain or in panic but as if he was just naturally ready to wake. This neural alarm clock works no matter if the mouse is in the dream-prone REM or the more physically restorative NREM sleep. Here, activa-

tion of neurons of the ventrolateral preoptic nucleus in the hypothalamus floods the brain with orexin. In alignment with the human literature, these same neurons are lost in people with narcolepsy, demonstrating that orexin acts as a wakefulness signal.

- Neuropathic pain in people can be debilitating. To mimic this disorder in mice, scientists do a surgical procedure that results in nerve injury. As in people living with chronic pain, these mice experience allodynia. Then the light turns on. The mouse experiences a significant alleviation of this abnormal pain. Instead of using ChR2, however, this manipulation uses a related light-sensitive protein called archaerhodopsin, which inhibits neurons. This archaerhodopsin has been put into sensory neurons that express the protein $Na_V1.8$, one of the proteins that allow for action potentials to travel down an axon.
- Humans and birds are pretty different, but language acquisition is one behavior that may have some shared evolutionary roots. Zebra finches, for example, learn a very stereotyped song made up of pitches and pauses that are like human syllables. Experimenters wait until the bird is in the middle of this song. Then the light turns on. The bird changes their song, singing the targeted syllable at a higher pitch and with more frequent repetitions. Dopaminergic terminals in the bird basal ganglia have been excited here, disrupting signaling through those structures used in habit learning.
- Rats are put onto a simple Y-shaped platform. At the end of one arm is a single pellet of rat chow. The other arm held either a whopping four pellets of food or a big fat nothing. Depending on the session, the odds of receiving the big reward are either high or low. This test is designed to mimic gambling and is called, fittingly, the two-arm bandit task.

> Rats are inherently decent at estimating probabilities over repeat trials, and the higher the odds of the big reward, the more often the rats would choose that wing. Then the light turns on. The rats transform from logical probability calculators into risk seekers. When inactivating the prelimbic cortex, rats preferentially gamble on the risky arm even if it means getting less total food over repeat trials. For these rats, the food itself might be secondary to the thrill of uncertainty, an experience sometimes analogous to what is seen in pathological gambling.

Dracula is so last century. In this day and age, we control our lab rodents with genetics and light.

> A merry little surge of electricity piped by automatic alarm from the mood organ beside his bed awakened Rick Deckard.
> —Philip K. Dick, *Do Androids Dream of Electric Sheep?*

In the opening scene of this cyberpunk classic, detective Deckard is awakened by a jolt of brain stimulation. His wife Iran does not wake with the same upbeat demeanor; Deckard takes a moment to mansplain that her settings were likely wrong. They alternate verbal blows, threatening to adjust the dial on their brain stimulators. Turn it to the left, and anger will dissipate. Turn it the other way: fireworks. In Philip K. Dick's universe, the Penfield mood organ can activate a precise pattern of neurons to evoke such specific moods as creativity, self-accusatory depression, and even a desire to sit and watch TV regardless of what is on. If this were in fact possible, it would be because of the pioneering work of Canadian neurosurgeon Wilder Penfield, after whom the device was named.

Penfield's specialty was epilepsy. Some types of seizures begin at a clump of nerve cells, called the focus. Once started, that electrical activity activates the connected cells akin to a row of dominoes falling sequentially. Severe seizures can be debilitating, especially if the person's muscles respond to the electrical activity violently, leading to falls or dislocated limbs. While modern seizure medications are relatively safe, the same couldn't be said of the medications of Penfield's day. The drug phenobarbital, when it hit the market in 1912, was decent for treating some cases of epilepsy, as it dampens neural activity and prevents the chain reaction of cellular excitation. Unfortunately, the drug had occasional terrible side effects. It led to fetal abnormalities, had the potential to be habit forming, and was even responsible for deaths since it sometimes shuts down the cells that are responsible for helping us breathe.

For patients with epilepsy, the alternative to a lifetime of risky medication was neurosurgery. If a surgeon could carefully extract the focus of the seizure activity, the condition could be improved if not completely cured. The difficulty is finding this clump of hyperactive tissue. An EEG could do the job, but since it sits on the surface of the head, it can approximate the location of only the strongest signals. This limitation can be overcome easily, at least if you have the creativity of a horror novelist and the brutality of a butcher. Why don't you just cut the skull open and attach the sensors directly against the surface of the brain? To get a precise reading of neuronal activity while performing open brain surgery, Penfield developed the electrocorticogram (ECoG), a set of probes small enough to rest on the surface of the brain. Compared to the EEG, the ECoG offers an increased ability to distinguish populations of cells, allowing surgeons to more accurately identify the offending area for targeted removal.

In addition to the many lives he improved through his scalpel skills, Penfield's long-lasting legacy was cemented by experiments he performed during those surgeries. After opening up his patient's skulls, he used a small electrical device to gently stimulate the surface of the brain while asking the patients to report their experiences. One part of the cortex caused a tingling sensation in the thumb, a nearby area caused the lower lip to tickle, and a third area caused the arm to contract. By thoroughly mapping patient experiences onto precise drawings of the brain, he became the first scientific* cortical cartographer, correlating parts of the brain with parts of the body. His brain–body map showed adjacent parallel strips of tissue running from the top of the brain down the side with two distinct functions. The one closer to the front of the brain led to muscle activity when stimulated, causing patients to blink, to make sucking motions with the mouth, or to point with the index finger. Some areas even caused vocalization, mostly grunting rather than anything intelligible. Because this strip of brain tissue leads to muscle group activity, it came to be called the motor cortex. Running electrical activity through the other strip of brain, the somatosensory cortex, caused people to report feeling sensations in different parts of their bodies, which they described as tingling, a coldness, a wave of flushing, numbness, and even feeling as though "it was going to sleep."

While electrically probing the human brain, Penfield discovered that the amount of brain tissue dedicated to certain body parts was not proportional to the size of that body part. For example, very little cortex was dedicated to the muscles and skin of the back, while much neural tissue controlled and sensed information from the hands. This lopsided brain–body correlation makes sense from a nervous system anatomical perspective. Compared to the rest of our body, the skin of the back has relatively few sensory nerves that detect sensations there. The skin of the hand, however, especially at the fingertips, is highly enriched in sensory nerves. This organizational discrepancy allows us to read Braille while we might struggle to locate something that is on our back. A bipedal representation of

* Even though they tried to tackle a similar goal, the lack of rigor disqualifies the phrenologists from this title.

the amount of brain dedicated to each body part yields a comically disproportionate figure, with a giant lower face, a shrunken body, spindly arms, and massive hands. This creature is Penfield's homunculus,* a word borrowed from ancient alchemical texts.

Stimulating other parts of the cortex produced all varieties of experiences. Some areas caused patients to see flashes of light, some induced smells, and yet others caused a strong desire to move a body part without any actual movement. Patients sometimes carried out a full complement of natural actions, such as complex tongue movements associated with swallowing, without consciously knowing what they were doing. Some stimulation initiated that frustrating, fleeting feeling of déjà vu. In one case, temporal lobe activation caused the patient to hear a song from the opera *Aida*.

Penfield is one of the heroes in the world of brain stimulation. He was given a moment of global recognition in the form of a Google Doodle on January 26, 2018, what would've been his 127th birthday. Depicted on the search engine's home page was an illustration of his face, a brain, and a piece of burnt toast, in honor of his patient who hallucinated the smell of that specific kitchen mishap during a brain stimulation experiment.

There is another figure who shared the same human brain stimulation space as Penfield but with a darker reputation. Robert Heath's main medical research interest was using brain stimulation to discover cures for mental illnesses, mostly with a focus on schizophrenia. Unlike Penfield, whose surgeries had been overall good for his patients, Heath's legacy was tarnished with a low success rate and side effects that were worse than what therapy might promise. For one woman, brain stimulation provoked a sudden display of fear, requiring five people to restrain her. For

* Although the critter is an exercise in body horror that crawled forth from the pits of the uncanny valley, people find him mostly endearing and goofy.

other patients, their side effects included a new history of seizures and fatal brain abscesses.

His most infamous study was conducted on a man, Patient B-19. This experiment was a study on the human reward circuitry. It had thus far been demonstrated that the septal region, a frontal lobe structure right along the midline, is part of the motivational circuitry. Burying a brain stimulator here and giving the participant a handheld button to stimulate these circuits gives the person unfettered access to nearly instant physical pleasure. Sometimes, people rub their thumbs raw from pushing the button, arguably hijacking a person's free will. In theory, septal self-stimulation may be useful for treating the anhedonia that accompanies schizophrenia, depression, anxiety, or PTSD. Whenever you feel a sense of despondency, panic, or an awful flashback coming on, just press the happiness button. It was an exercise in classical Pavlovian conditioning, with the goal of rewiring associations between some stimulus and the feeling of pleasure.

Heath tried to reprogram a homosexual man. After B-19 was surgically implanted with septal electrodes, a female sex worker was recruited and brought into the laboratory. The two of them were given a private space, and septal stimulation was delivered at the same time as bodily stimulation. After the experiment, according to Heath's reports, B-19 no longer had a desire to be with men, having engaged in gay sex twice only when he needed money. As further evidence of the "success" of this brain manipulation, B-19 initiated a 10-month-long sexual relationship with a married woman. At the time of these studies, homosexuality was considered a paraphilia according to the *Diagnostic and Statistical Manual of Mental Disorders*. But today, we look back on this application of the procedure with the same disgust as the ice-pick lobotomy and insulin-induced coma therapy.*

* Heath's list of unethical research included likely data fabrication that a protein called taraxein causes schizophrenia, followed by the subsequent injection of that protein into incarcerated people.

Both Penfield's and Heath's brain stimulation operations, whether for good or for bad, require open-head surgery, a dangerous process with tremendous risks for the patient. It would be much more advantageous and more likely to be adopted were it possible to do the stimulation without having to go under the scalpel. And that's the thought process that guided scientists to refine noninvasive methods for brain stimulation, such as transcranial direct current stimulation and transcranial magnetic stimulation, the latter of which changes brain activity using the magic of physics.* Although there isn't a clear consensus among the different regulatory agencies around the world, brain stimulation tools have been approved in some places for therapy of complex neuropsychiatric conditions, such as major depressive disorder, OCD, substance use disorder, anxiety, and migraine pain. In preclinical studies, brain stimulation has also been used for neurorehabilitation, especially for the cognitive deficits seen after a stroke.

In addition to these applications, there are other, more subtle effects of brain stimulation for healthy people, but the improvements are modest. You might be aware of a cognitive bias called the negativity effect (if not by name, then by experience). We have a natural tendency to dwell on unpleasant events compared to equally strong positive events.† There is a related bias that goes in the opposite direction, called the good news/bad news effect. In this case, we are more likely to remember good news about our future selves compared to bad news. Again, there is an asymmetry in how well we remember things but this time toward optimism.

For example, make a mental guess right now. What do you suppose are the odds you will have a stroke in your lifetime? Consid-

* Arthur C. Clarke's third law applies here. "Any sufficiently advanced technology is indistinguishable from magic." I'm not above admitting that principles in physics such as induction are basically magic to me.

† I have no idea how many times that I, as an elementary school child, won a game of four square, did not get picked last for dodgeball, or aced a math quiz. But I can vividly recall the reactions I got the day I showed up to school with my face framed by purple orthodontic headgear.

ering how many people you might know personally, how often you see stroke as a side effect of a medication, or how many celebrities survive a stroke, you might guess your risk is as high as the double digits. Luckily, here's the good news: the risk isn't nearly as high as you thought. It's closer to 3 percent. Considering this good news, if you were asked to make predictions about stroke risk in the future, you might guess a number in the single digits, possibly even accurately recalling the 3 percent risk. Here's a second condition. What do you suppose are the odds you will develop Alzheimer's disease between the ages of 75 and 84? If your guess is anything less than 10 percent, I have bad news.* This time, your risk is much closer to 13 percent. If you were asked about this number in the future, you are less likely to remember just how common the disease is. We selectively update our knowledge in light of positive facts, but we struggle to remember where we put our half-empty glass.

With the right pattern of transcranial magnetic brain stimulation, this learning bias can be overcome. The left inferior frontal gyrus plays an important role in this specific type of relearning, so activating these circuits helps people look on the pessimistic side of things. But who, other than Decker's wife, who voluntarily set her brain stimulator to "self-accusatory depression," actually wants that? For one, these studies demonstrate that memories can be improved, which is an outcome that scientists have long desired. Second, being able to better predict negative outcomes can help guide policymakers toward preparing realistically for natural disasters by focusing on the worst possible scenario. It can help prevent financial experts from making poor macroeconomic decisions during financial bubbles, better utilizing past experiences to guide

* Prevalence statistics are always a bit fuzzy. There are many risk factors that influence these numbers, including demographics (women are about twice as likely to have an Alzheimer's disease diagnosis than men), lifestyle choices (physical inactivity increases risk for stroke), environmental influences (pollution exposure increases risk for both outcomes), and, of course, how likely it is that a person even goes to get a diagnosis in the first place.

future plans. It may also help physicians more accurately explain the risks of treatments to their patients.

Brain stimulation can even change our perceptions of other people. Transcranial stimulating electrodes were placed on the scalp right along the midline of a handful of Dutch nationals. Then they were asked to perform the implicit association task* to reveal if they held biases against Moroccan people, a large and growing immigrant group in the Netherlands. During brain stimulation, participants answered quicker and more accurately on the incongruent trials, those that asked the participant to pair names such as Habib and Salim with positive words such as *vrede* (peace) and *gelukkig* (happiness).

Before you consider performing amateur brain stimulation on yourself for a tiny bit of self-improvement on memory or language tasks, be aware that the wrong kind of transcranial stimulation, maybe because it was targeted to a different area or it fired off an unnatural pattern of activity, can work against you. Instead of achieving the desired result of turning up the electrical activity of a certain circuit, the stimulator might act more like a temporary lesion, shutting down the activity of those neurons. Most likely, the healthy brain is already operating at peak performance—that is, there's not much upwards room for improvement. Granted, these noninvasive stimulation tools don't yet offer precise targeting of neurons in the same way a genetic modification such as optogenetics can. Brain stimulation is still a crude sledgehammer of a tool, a far cry from the workings of Philip K. Dick's Penfield mood organ.

* If the details of this task are fuzzy, try a little neurostimulation of the hippocampus, which has been shown to improve performance on declarative memory tasks. Alternatively, check chapter 10.

CLOSING COMMENTS

THIS BOOK ENDS THE SAME WAY THAT SCIENCE FICTION AND horror began: *Frankenstein*.

It is easy to view *Frankenstein* as a precautionary tale about the dangers of obsession. Initially, the reader gets the impression that young Victor is acting through curiosity, that intrinsic force that brought you to the end of the book and the trait that doomed humanity since the "Adam and Eve eat the fruit of the tree of Good and Evil" creation myth. Throughout the story, we see flashes of Victor's obsession, twisted through the lens of loss and revenge, with his final obsession being that of destroying his own creation.

While Victor's obsession brought him to a terrible place, that trait is not why he is a tragic character. His downfall is that his unquenchable desire robbed him of his humanity, no longer able to empathize with the creation he birthed. He forgot that his creation was made from human beings. He was blinded to its loneliness, and in refusing to create a partner for the creature, he denied it the possibility of forming a meaningful connection, the one thing it sought more than anything. The creature was not a monster. It was the man, Dr. Frankenstein, who lost touch with his humanity, who was the monster.

Here's the challenge that I issue for scientists, young and experienced alike, whose research goals are noble but abstract: seek the humanity in whatever you do.

My love of science was the thrill of discovery. It's an irreplaceable feeling, one that has nearly no parallel outside of the laboratory. There's a moment in time when you are the only person in the world to know something. I imagine it's the feeling that drives Indiana Jones, an occasional action hero who pays his bills by working a day job as a college professor, to plunder ruined temples for mysterious artifacts. It's the same reason why characters in horror films voluntarily enter the undisputedly haunted house, why Lovecraftian protagonists keep turning the pages of forbidden tomes, and why exploring exotic worlds is so captivating: massively different questions but the exact same energy.

There's a nearsightedness that creeps up on research scientists like me who get enthralled with discovery, especially when those findings are tiny and incremental, the way most are these days. We get so deeply immersed with protein-protein complexes, neuronal membrane potentials, or computer models that we lose track of the end goal. The day-to-day optics are simply too zoomed in. Without a doubt, sharing knowledge of how neurons in a culture dish respond to opioids is important. But it is also immensely meaningful to step away from the clean room and volunteer at a rehabilitation clinic or a needle exchange program. When you're done quantifying the memory skills of genetically modified Alzheimer's disease–like mice, get a caregiver license, become a certified dementia practitioner, or work to reduce personal stigma through regularly interacting with patient support groups.

I was caught completely off guard by my own personal "put a face on the science" moment. Most of the data described in my PhD thesis were cellular data collected from neurons in a thin slice of mouse brain tissue, extracted from the animal but maintained alive in an artificial bath of salts and nutrients. The big picture was to understand the cellular mechanisms of levodopa-induced dyskinesia, a disorder that causes uncontrolled movements in people with Parkinson's disease who are taking medication. It's a side effect that becomes more severe the longer the person takes their medication and is nearly the inevitable fate of patients on the drug:

up to of 90 percent of patients develop dyskinesias after 15 years of treatment. Ironically, these uncontrolled movements can be just as debilitating as Parkinson's disease, the reason why these patients take the medication to begin with.

Each year, the neuroscience community congregates at Society for Neuroscience, a massive international conference that attracts some 40,000 participants. This five-day-long event is crammed full of poster presentations detailing hot-off-the-press data, talks by the most influential leaders, panels of experts debating controversies in the field, and robust networking opportunities, often by means of an inordinate amount of drinking alongside some of the most respected professors. The breadth of topics covered is impressive; posters are usually organized thematically by such specific banners as "Human Medial Temporal Lobe" and "Neuroinflammation: Zika." The poster session takes place in a massive convention hall, and posters are assigned an alphanumeric code. You're just as likely to find interesting data at poster P12 as you are at poster FFF29—triple letters being the next resort after the first 26, then the next 26 of double letters, have been assigned.

As a graduate student, I attended a handful of these conferences to present some tiny fragments of an incomplete project, mostly for the purpose of learning how to organize results and share scientific data. At one such conference, I stood in front of my poster, dehydrated and a little queasy in the wake of a Bourbon Street–fueled night of mystery shots courtesy of the infamous "Dopamine Dinner," an unsanctioned gathering where, for the time being, scientists were concerned more with the firsthand effects of dopamine signaling. In between a steady flow of grad students, senior professors, and everyone in between, I was approached by two people: an older gentleman and a woman. Many attendees stroll past the posters, glancing at the titles or the images until they come across something they are curious to learn more about, perhaps landing on some tiny piece of data that could be of use in their own research project. The man stopped in front of my poster, looking curiously over my recordings of action potentials and poorly drawn diagrams of neurons. I described the data on the poster, giving my canned introduction that ties the clinical question of dyskinesia to neurons while glossing over a lot of information about neurotransmitter release and receptor proteins. The spiel was well rehearsed by this point, having given it to so many fellow neuroscientists who had stopped by the poster throughout the morning. After describing my conclusions and my next planned experiments, I asked if there were any questions I could address. He smiled, introduced himself to me, and explained his interest in my poster.

He is living with Parkinson's disease on a regimen of levodopa for his symptoms. He knew about dyskinesia not from dry clinical descriptions or academic publications but from firsthand, daily lived experience. He was perusing the half-mile-long conference hall, walking alongside his caretaker, looking for potential home remedies or any scrap of knowledge he could bring to his physician for treating his dyskinesia. I quickly backpedaled the data on my poster, stumbling through all the caveats: that the data were tested in an experimental mouse model of Parkinson's disease, that the affected neurons are just one part of many interacting circuits, that levodopa is given to the mice differently than people, and so on. I merely found that one type of neuron behaves differently when chronically exposed to levodopa. My poster didn't outline a treatment protocol for dyskinesia. The last thing I wanted to do was to deceive this man or to give him any false hope.

After all this, he kindly smiled again, thanked me, and then left me with a bit of encouragement that has shifted my mindset about research ever since.

"Well, I hope you find what you are looking for."

Normally, chats in front of the poster are initiated by fellow scientists, candidly spitballing thoughts and half-baked ideas for new experiments pulled from an individualized lens through which they view the process of research. This man, however, had given me something much more valuable than an idea. He realigned my purpose for doing science. The goal suddenly became to discover how to help this one man in particular. Sure, I had seen low-resolution, grainy video clips of patients battling their dyskinesia in clinical examinations. But there is a huge detachment between graduate student me and the person struggling their hardest to stay still in a chair. For the first moment in my time doing bench science, patients with dyskinesia were more than just a statistic, more than just a sentence in a manuscript's introduction. This one encounter transformed the ambiguous "patients" into a human, someone whose hand I can shake.

That was the moment when my curiosity merged with my humanity. And I wish every scientist could have these moments.

Once you've had this experience, go a mile farther. With that one person in your mind, think about the hundreds, thousands, or maybe millions more just like them. Start with Society for Neuroscience's advocacy training. Take that knowledge, then lobby for policymakers to pass legislation that benefits those people. Push local politicians to develop infrastructure in accordance with the World Health Organization's age-friendly cities initiative (after all, the 2020s is the United Nations Decade of Healthy Ageing). If you are in a position of power at your institution, encourage the creation of training programs for scientists who want to work in public policy and emphasize the importance of this work.

What can advocacy actually do? Consider Henri Emmanueli, a member of the French Parliament for the department of Landes in southwestern France. With the equivalent of $22 million and the support of his constituency, he established the Village Landais, a dementia-friendly community more colloquially known as "Alzheimer's Village." Here, 120 patients live their fullest lives independently, spending their time walking through parks and gardens, shopping sans wallets at cafés and restaurants, and enjoying the benefits of a library, an auditorium, and even a farm with animals (the villagers are particularly proud of the two donkeys). The one thing it doesn't have is white coats: although there are several medical doctors, therapists, and psychogeriatricians interspersed among the employees, any obvious signs of medicalization have been intentionally stripped away and, along with it, the fears and anxieties that accompany medical visits. As it opened in 2020, it is relatively young, so the long-term benefits for the patients haven't been published yet, although preliminary evidence suggests patients displayed a maintenance in cognitive performance and quality of life, a major accomplishment to stem the tide of destruction brought about by a progressive neurodegenerative disease. There are also clear benefits for the people living in the surrounding town. Typically, people with Alzheimer's disease are perceived as being low skilled, leading to stigmatization and exclusion, leading to worsened health outcomes. Since the opening of Alzheimer's Village, residents of the neighboring town reported that patients with Alzheimer's disease were more capable. This initiative is just one example of how advocacy can bring humanity to science. Hogeweyk Village in the Netherlands features different backdrops among which patients can live, such as an urban setting, one with a strong religious focus, and even one reminiscent of Indonesia. Its construction was paid for largely by funds from the Dutch government. The first of such care centers in the United States, Glenner Town Square in San Diego, is styled to resemble an average Main Street from the 1950s, complete with public telephone booths and *I Love Lucy* on the televisions. Dr. Glenner is a major advocate on behalf of patients, having encouraged Ronald Reagan to designate November as Alzheimer's Awareness Month, which in turn led to an increase in research funding from the National Institutes of Health. Advocacy helps both patients and scientists.

Think back to the horror posed in the introduction suggesting that all behavior is a consequence of neurons releasing chemicals.

The more nuanced observation, as revealed by the intervening pages, is that those neurons are under the control of genes and that their transcription into protein is regulated by environmental influences. Our fate is not sealed from the moment the cells wire themselves; studies have shown repeatedly that even identical twins have differing rates of neurodegenerative diseases, depression, anxiety, and other measures of brain health. We have the capacity to influence how those genes make protein and hence neuronal activity by changing the surroundings that people live in.

But most people do not have the means to build an environment for optimal brain development or health. In rats, environmental enrichment increases synaptogenesis, quickens recovery after brain injury, decreases drug seeking and relapse after abstinence, and many other benefits. These findings probably map onto humans too. A fully enriched environment for people accounts for every stage, starting with prenatal maternal care, accessible day care for children, fulfilling after school programs, public libraries, health care and counseling services, outdoor exercise spaces, support services for seniors, and end-of-life care.

Unfortunately, politicians at the local and national levels have actively sought to defund all those initiatives, using whatever hot topic issue they can twist to fit the narrative that will best leverage their voter base or appease their campaign donors. They have repeatedly worked to remove access to mental health services and educational opportunities, to criminalize safe injection sites, and to roll back environmental protection regulations on organophosphate insecticides, a known risk factor for neurodegenerative diseases. They have made it more costly to access medications and weakened the infrastructure of financial safety nets, thereby increasing the strain on impoverished people. Outside of their voting record, they have publicly mocked academic research, worsening public opinion about science by devaluing the people who do the work.

The real horror of our existence is not that we are controlled by our neurons but that we are controlled by others. We understand what is needed to improve conditions for humanity, but other humans actively stand in the way. In this roundabout tour through the nervous system, I arrived at the same conclusion that Stephen King did in the last chapter of *Danse Macabre*, his retrospective account of the history of horror fiction. Interspersed throughout the chapter are anecdotes of real-life horrific murders, from stabbings to immolation, "inspired" by the killings depicted in fiction. The scariest thing out there is not a ghost in the well or a clown in the storm drain. It's other people.

I hope to leave you with this tiny call to action. Become obsessed with your humanity and seeking humanity in others. Create things, stories, and knowledge. Connect with people the way only humans can. Care for others and inspire them. Wonder about the world. Do those good human things. You are more than just a brain.

It is the goal of all to improve, advance, progress, grow. —Bella Baxter, *Poor Things* (2023)

ACKNOWLEDGMENTS

MY LOVE OF SCIENCE COMES FROM THE SOLID FOUNDATIONS built by the educators of my childhood. Mrs. Brock, my first and second grade teacher, gave me the freedom to explore. Mr. Mathews, my fourth and fifth grade science teacher, kindled my curiosity with his lab demonstrations. Mrs. Wheeler encouraged me to start journaling to get over a fear of writing. And my high school speech team coaches, Mr. and Mrs. Etherton, cultivated both my confidence and my communication skills.

I'm sincerely grateful for my previous academic advisers, all of whom took some sort of chance on me. Dr. Tony West at Rosalind Franklin University had little to gain from taking a political science major into his rat pharmacology lab, but he did anyway. My thesis adviser, Dr. Dan McGehee at the University of Chicago, recognized my passion for neuroscience despite my shallow pool of actual knowledge. After a fruitful collaborative project with Dr. Un Jang Kang, one first-class plane ticket, and some years of guidance later, I finished my PhD. Then my postdoctoral adviser, Dr. Jim Surmeier at Northwestern University, hired me despite my job talk that showed data contradicting some of his previous findings. Looking back at my time under his leadership, I didn't take full advantage of the plethora of tools available to me. But being surrounded by an international team of talented and dedicated neuroscientists working on cutting-edge research was one of the most exciting times in my career. I hope that I can inspire others the same way these three leaders have inspired me.

ACKNOWLEDGMENTS

Many of the stories in this book were collected over the years of preparing lectures for my students at DePaul University. I am proud that my students, especially those in my "Brain through Science Fiction" course, pose some of the most thought-provoking questions, exposing the gaps in my knowledge that I would have never thought to fill. My coworkers in the newly minted Department of Neuroscience are the most supportive and helpful fellow professors I could ask for, and I am honored to work alongside you all.

Of course, thank you to all the scientists who conducted the research on which this book is built. You are the underappreciated giants on which others stand. On the other side of the equation, there are also many flavors of artists, authors, comic book creators, filmmakers, screenwriters, and more who created our favorite and most feared characters and settings. Thanks for inspiring my imagination.

This book wouldn't be in your hands without the work of my agent, Lane Heymont, at the Tobias Literary Agency. You've been so generous with your time and an excellent advocate on my behalf. Thank you to my editors (Jake Bonar, Felicity Tucker, and Bruce Owens) at Prometheus Books and Globe Pequot for helping refine my initial vision into this book.

I also want to thank my parents, Allen and Sue. Over the years, you have given me unconditional love and encouragement without which this book could not exist. The sacrifices you have made for me are appreciated every single moment. You instilled in me the values of dedication and hard work, values that I hope to also pass along to my own children.

And to my family: Allie, Polo, and Aldie. You inspire me to be a better man every day. You are the heart of my humanity. I love you.

NOTES

INTRODUCTION

vii **Modern neuroanatomists estimate that:** Suzana Herculano-Houzel, "The Human Brain in Numbers: A Linearly Scaled-Up Primate Brain," *Frontiers in Human Neuroscience* 3 (2009): 31, https://doi.org/10.3389/neuro.09.031.2009.

vii **a single neuron might connect with more than a few hundred thousand other neurons:** Toby Tyrrell and David Willshaw, "Cerebellar Cortex: Its Simulation and the Relevance of Marr's Theory," *Philosophical Transactions of the Royal Society of London. Series B: Biological Sciences* 336, no. 1277 (1997): 239–57, https://doi.org/10.1098/rstb.1992.0059.

ix **the storage of data happens not only at the level of the synapse:** Yingxu Wang, Dong Liu, and Ying Wang, "Discovering the Capacity of Human Memory," *Brain and Mind* 4, no. 2 (2003): 189–98, https://doi.org/10.1023/A:1025405628479.

xi **why a measurement of hemodynamic response through functional magnetic resonance imaging (fMRI) is not a true assay of how neuronal firing changes:** Aaron T. Winder, Christina Echagarruga, Qingguang Zhang, and Patrick J. Drew, "Weak Correlations between Hemodynamic Signals and Ongoing Neural Activity during the Resting State," *Nature Neuroscience* 20, no. 12 (2017): 1761–69, https://doi.org/10.1038/s41593-017-0007-y.

xii **why it is so difficult to use genetically modified mice to learn about the human condition:** Klaus I. Matthaei, "Genetically Manipulated Mice: A Powerful Tool with Unsuspected Caveats," *Journal of Physiology* 582, no. 2 (2007): 481–88, https://doi.org/10.1113/jphysiol.2007.134908.

CHAPTER 1

4 **farther upstream of the hypothalamus is the amygdala:** Elizabeth R. Duval, Arash Javanbakht, and Israel Liberzon, "Neural Circuits in Anxiety and Stress Disorders: A Focused Review," *Therapeutics and Clinical Risk Management* 11 (2015): 115–26, https://doi.org/10.2147/TCRM.S48528.

4 **cells of the pulvinar seem to be particularly sensitive to snakes:** Quan Van Le, Lynne A. Isbell, Jumpei Matsumoto, Minh Nguyen, Etsuro Hori, Rafael S. Maior, Carlos Tomaz, Anh Hai Tran, Taketoshi Ono, and Hisao Nishijo, "Pulvinar Neurons Reveal Neurobiological Evidence of Past Selection for Rapid Detection of Snakes," *Proceedings of the National Academy of Sciences of the United States of America* 110, no. 47 (2013): 19000–19005, https://doi.org/10.1073/pnas.1312648110.

5 **The insula . . . cares a great deal about disgust:** Xianyang Gan, Feng Zhou, Ting Xu, Xiaobo Liu, Ran Zhang, Zihao Zheng, Xi Yang, X. Zhou, F. Yu, J. Li, R. Cui, L. Wang, J. Yuan, D. Yao, and B. Becker, "A Neurofunctional Signature of Subjective Disgust Generalizes to Oral Distaste and Socio-Moral Contexts," *Nature Human Behaviour* 8 (April 2004): 1383–1402, https://doi.org/10.1038/s41562-024-01868-x.

7 **SM scored poorly on identifying them as being scared:** R. Adolphs, D. Tranel, H. Damasio, and A. Damasio, "Impaired Recognition of Emotion in Facial Expressions Following Bilateral Damage to the Human Amygdala," *Nature* 372, no. 6507 (1994): 669–72, https://doi.org/10.1038/372669a0.

8 **being held at knifepoint:** Justin S. Feinstein, Ralph Adolphs, Antonio Damasio, and Daniel Tranel, "The Human Amygdala and the Induction and Experience of Fear," *Current Biology* 21, no. 1 (2011): 34–38, https://doi.org/10.1016/j.cub.2010.11.042.

8 **forced to breathe through a face mask supplied with air containing 35 percent carbon dioxide:** Justin S. Feinstein, Colin Buzza, Rene Hurlemann, Robin L. Follmer, Nader S. Dahdaleh, William H. Coryell, Michael J. Welsh, Daniel Tranel, and John A. Wemmie, "Fear and Panic in Humans with Bilateral Amygdala Damage," *Nature Neuroscience* 16, no. 3 (2013): 270–72, https://doi.org/10.1038/nn.3323.

11 **at least one person has died because of an oxygen tank being accelerated:** David W. Chen, "Boy, 6, Dies of Skull Injury during M.R.I. (Published 2001)," *New York Times*, July 31, 2001, sec. New York, https://www.nytimes.com/2001/07/31/nyregion/boy-6-dies-of-skull-injury-during-mri.html.

12 **pictures of spiders:** Christine L. Larson, Hillary S. Schaefer, Greg J. Siegle, Cory A. B. Jackson, Michael J. Anderle, and Richard J. Davidson, "Fear Is Fast in Phobic Individuals: Amygdala Activation in Response to Fear-Relevant Stimuli," *Biological Psychiatry* 60, no. 4 (2006): 410–17, https://doi.org/10.1016/j.biopsych.2006.03.079.

12 **digital avatar pursued by a virtual predator:** Song Qi, Demis Hassabis, Jiayin Sun, Fangjian Guo, Nathaniel Daw, and Dean Mobbs, "How Cognitive and Reactive Fear Circuits Optimize Escape Decisions in Humans," *Proceedings of the National Academy of Sciences of the United States of America* 115, no. 12 (2018): 3186–91, https://doi.org/10.1073/pnas.1712314115.

12 **watching two of the top 10 scariest movies:** Matthew Hudson, Kerttu Seppälä, Vesa Putkinen, Lihua Sun, Enrico Glerean, Tomi Karjalainen, Henry K. Karlsson, Jussi Hirvonen, and Lauri Nummenmaa, "Dissociable Neural Systems for Unconditioned Acute and Sustained Fear," *NeuroImage* 216 (August 2020): 116522, https://doi.org/10.1016/j.neuroimage.2020.116522.

12 **amygdala activity decreases under brighter light conditions:** Elise M. McGlashan, Govinda R. Poudel, Sharna D. Jamadar, Andrew J. K. Phillips, and Sean W. Cain, "Afraid of the Dark: Light Acutely Suppresses Activity in the Human Amygdala," *PLoS One* 16, no. 6 (2021): e0252350, https://doi.org/10.1371/journal.pone.0252350.

12 **the signal in the subgenual anterior cingulate cortex increased:** Uri Nili, Hagar Goldberg, Abraham Weizman, and Yadin Dudai, "Fear Thou Not: Activity of Frontal and Temporal Circuits in Moments of Real-Life Courage," *Neuron* 66, no. 6 (2010): 949–62, https://doi.org/10.1016/j.neuron.2010.06.009.

13 **Mori plotted a theoretical graph:** Masahiro Mori, "Bukimi no tani [The Uncanny Valley]," *Energy* 7 (1970): 33.

16 **some of them started to cry:** NPR.org, "Hollywood Eyes Uncanny Valley in Animation," n.d., https://www.npr.org/templates/story/story.php?storyId=124371580.

17 **modeled after a genuine medical condition:** Cara Warner, "Lupita Nyong'o Explains How She Came Up with Her *Us* Movie Character's Super Creepy Voice," *People*, March 23, 2019, https://people.com/movies/lupita-nyongo-explains-us-character-creepy-voice.

17 **visual stimuli can also initiate a disgust response:** Mahdi Muhammad Moosa and S. M. Minhaz Ud-Dean, "Danger Avoidance: An Evolutionary Explanation of Uncanny Valley," *Biological Theory* 5, no. 1 (2010): 12–14, https://doi.org/10.1162/BIOT_a_00016.

18 **mortality salience:** Karl F. MacDorman and Hiroshi Ishiguro, "The Uncanny Advantage of Using Androids in Cognitive and Social Science Research," *Interaction Studies: Social Behaviour and Communication in Biological and Artificial Systems* 7, no. 3 (2006): 297–337.

19 **signals in this area decrease when we are asked to rate the likability of uncanny residents:** Astrid M. Rosenthal-von der Pütten, Nicole C. Krämer, Stefan Maderwald, Matthias Brand, and Fabian Grabenhorst, "Neural Mechanisms for Accepting and Rejecting Artificial Social Partners in the Uncanny Valley," *Journal of Neuroscience* 39, no. 33 (2019): 6555, https://doi.org/10.1523/JNEUROSCI.2956-18.2019.

19 **The vmPFC communicates strongly with the amygdala:** Jaryd Hiser and Michael Koenigs, "The Multifaceted Role of the Ventromedial Prefrontal Cortex in Emotion, Decision Making, Social Cognition, and Psychopathology," *Biological Psychiatry* 83, no. 8 (2018): 638–47, https://doi.org/10.1016/j.biopsych.2017.10.030.

19 **When shown a continuum of faces:** Marcus Cheetham, Pascal Suter, and Lutz Jäncke, "The Human Likeness Dimension of the 'Uncanny Valley Hypothesis': Behavioral and Functional MRI Findings," *Frontiers in Human Neuroscience* 5 (2011), https://www.frontiersin.org/journals/human-neuroscience/articles/10.3389/fnhum.2011.00126.

22 **responds fairly selectively for images of houses:** R. Epstein and N. Kanwisher, "A Cortical Representation of the Local Visual Environment," *Nature* 392, no. 6676 (1998): 598–601, https://doi.org/10.1038/33402.

22 **increases its signal to similar stimuli:** Moshe Bar and Elissa Aminoff, "Cortical Analysis of Visual Context," *Neuron* 38, no. 2 (2003): 347–58, https://doi.org/10.1016/s0896-6273(03)00167-3.

22 **miniature room built out of Lego bricks:** Thomas Wolbers, Roberta L. Klatzky, Jack M. Loomis, Magdalena G. Wutte, and Nicholas A. Giudice, "Modality-Independent Coding of Spatial Layout in the Human Brain," *Current Biology* 21. no. 11 (2011): 984–89, https://doi.org/10.1016/j.cub.2011.04.038.

22 **It likely also encodes boundary information:** Frederik S. Kamps, Joshua B. Julian, Jonas Kubilius, Nancy Kanwisher, and Daniel D. Dilks, "The Occipital Place Area Represents the Local Elements of Scenes," *NeuroImage* 132 (May 2016): 417–24, https://doi.org/10.1016/j.neuroimage.2016.02.062.

CHAPTER 2

25 **Swammerdam and his rigorous science:** John Pearn, "A Curious Experiment: The Paradigm Switch from Observation and Speculation to Experimentation, in the Understanding of Neuromuscular Function and Disease," *Neuromuscular Disorders* 12, no. 6 (2002): 600–607, https://doi.org/10.1016/S0960-8966(01)00310-8.

25 **developed a preparation consisting of an airtight glass vacuum tube:** M. Cobb, "Timeline: Exorcizing the Animal Spirits: Jan Swammerdam on Nerve Function," *Nature Reviews Neuroscience* 3, no. 5 (2002): 395–400, https://doi.org/10.1038/nrn806.

26 **Galvani would develop:** Marco Piccolino, "Luigi Galvani's Path to Animal Electricity," *Aspects de l'histoire Des Neurosciences* 329, no. 5 (2006): 303–18. https://doi.org/10.1016/j.crvi.2006.03.002.

26 **Aldini, continued in his uncle's footsteps:** "Account of Late Improvements in Galvanism, by John Aldini—A Project Gutenberg EBook," n.d., https://www.gutenberg.org/files/57267/57267-h/57267-h.htm.

28 **jaws of the deceased criminal:** André Parent, "Giovanni Aldini: From Animal Electricity to Human Brain Stimulation." *Canadian Journal of Neurological Sciences* 31, no. 4 (2004): 576–84, https://doi.org/10.1017/s0317167100003851.

29 **this giant axon is half a millimeter:** Christof J. Schwiening, "A Brief Historical Perspective: Hodgkin and Huxley," *Journal of Physiology* 590, no. 11 (2012): 2571–75, https://doi.org/10.1113/jphysiol.2012.230458.

31 **vagus nerve activation leads to:** Ying Li and Chung Owyang, "Musings on the Wanderer: What's New in Our Understanding of Vago-Vagal Reflexes? V. Remodeling of Vagus and Enteric Neural Circuitry after Vagal Injury," *American Journal of Physiology. Gastrointestinal and Liver Physiology* 285, no. 3 (2003): G461–69, https://doi.org/10.1152/ajpgi.00119.2003.

31 **these nerves function even if the vagus nerve is severed:** W. M. Bayliss and E. H. Starling, "The Movements and Innervation of the Small Intestine," *Journal of Physiology* 24, no. 2 (1899): 99–143, https://doi.org/10.1113/jphysiol.1899.sp000752.

31 **came to him in a dream:** Alli N. McCoy and Siang Yong Tan, "Otto Loewi (1873–1961): Dreamer and Nobel Laureate," *Singapore Medical Journal* 55, no. 1 (2014): 3–4, https://doi.org/10.11622/smedj.2014002.

32 **acetylcholine is conserved:** Yoko Horiuchi, Reika Kimura, Noriko Kato, Takeshi Fujii, Masako Seki, Toyoshige Endo, Takashi Kato, and Koichiro Kawashima, "Evolutional Study on Acetylcholine Expression," *Life Sciences* 72, no. 15 (2003): 1745–56, https://doi.org/10.1016/S0024-3205(02)02478-5.

34 **Dr. Christiaan Barnard performed:** David K. C. Cooper, "Christiaan Barnard—The Surgeon Who Dared: The Story of the First Human-to-Human Heart Transplant," *Global Cardiology Science and Practice* 2018, , no. 2 (2018): 11, https://doi.org/10.21542/gcsp.2018.11.

35 **more than 4,000 times:** United Network for Organ Sharing, "UNOS Data and Transplant Statistics | Organ Donation Data," 2023, https://unos.org/data.

36 **hand transplant procedure has been performed:** Maria João Lúcio and Ricardo Horta, "Hand Transplantation-Risks and Benefits," *Journal of Hand and Microsurgery* 13, no. 4 (2021): 207–15, https://doi.org/10.1055/s-0040-1715427.

36 **Like Luke Skywalker:** J. M. Dubernard, E. Owen, G. Herzberg, M. Lanzetta, X. Martin, H. Kapila, M. Dawahra, and N. S. Hakim, "Human Hand Allograft: Report on First 6 Months," *Lancet* 353, no. 9161 (1999): 1315–20, https://doi.org/10.1016/S0140-6736(99)02062-0.

36 **considering that more than 100,000 Americans:** Health Resources and Services Administration, "Organ Donation Statistics," October 2023, https://www.organdonor.gov/learn/organ-donation-statistics.

36 **first pig cornea:** R. Kissam, "Ceratoplastice in Man," *New York Journal of Medicine* 2 (1844): 281–82.

36 **Of the 13 human recipients . . . sections of baboon testicles:** David K. C. Cooper, Burcin Ekser, and A. Joseph Tector, "A Brief History of Clinical Xenotransplantation," *International Journal of Surgery* 23, pt. B (2015): 205–10, https://doi.org/10.1016/j.ijsu.2015.06.060.

37 **survived only two months:** Associated Press, "A Man Who Got the 1st Pig Heart Transplant Has Died after 2 Months," NPR, March 9, 2022, sec. Health, https://www.npr.org/2022/03/09/1085420836/pig-heart-transplant.

37 **Some wonder:** Kuheli Biswas, "Recipient of Genetically Modified Pig Heart Was Convicted in 80s Stabbing Case," *International Business Times*, January 14, 2022, https://www.ibtimes.com/historic-recipient-genetically-modified-pig-heart-was-convicted-80s-stabbing-case-3375506.

38 **a radical surgery proposed:** Sergio Canavero, "HEAVEN: The Head Anastomosis Venture Project Outline for the First Human Head Transplantation with Spinal Linkage (GEMINI)," *Surgical Neurology International* 4, suppl. 1 (2013): S335–42, https://doi.org/10.4103/2152-7806.113444.

39 **Ren's spinal cord repair procedure:** Zehan Liu, Shuai Ren, Kuang Fu, Qiong Wu, Jun Wu, Liting Hou, Hong Pan, Linlin Sun, Jian Zhang, Bingjian Wang, Qing Miao, Guiyin Sun, Vincenzo Bonicalzi, Sergio Canavero, and Xiaoping Ren, "Restoration of Motor Function after Operative Reconstruction of the Acutely Transected Spinal Cord in the Canine Model," *Surgery* 163, no. 5 (2018): 976–83, https://doi.org/10.1016/j.surg.2017.10.015.

39 **rats died in a flood:** C.-Yoon Kim, William K. A. Sikkema, In-Kyu Hwang, Hanseul Oh, Un Jeng Kim, Bae Hwan Lee, and James M. Tour, "Spinal Cord Fusion with PEG-GNRs (TexasPEG): Neurophysiological Recovery in 24 Hours in Rats," *Surgical Neurology International* 7, suppl. 24 (2016): S632–36, https://doi.org/10.4103/2152-7806.190475.

40 **the total cost of the operation:** Kim Hjelmgaard, "Head Transplant Doctors Xiaoping Ren and Sergio Canavero Claim Spinal Cord Progress," *USA Today*, March 27, 2019, https://www.usatoday.com/story/news/world/2019/03/27/italian-chinese-surgeons-cite-spinal-cord-repair-head-transplant-canavero-xiaoping/3287179002.

40 **He insists that his pioneering work:** "Scientists Sound Warnings about Man Who Says He Is About to Perform a Head Transplant," *The Independent*, November 17, 2017, https://www.independent.co.uk/news/science/head-transplant-surgery-sergio-canavero-body-medical-reaction-criticism-real-a8061011.html.

CHAPTER 3

45 **A brush with death:** Rümeysa İnce, Saliha Seda Adanır, and Fatma Sevmez, "The Inventor of Electroencephalography (EEG): Hans Berger (1873–1941)," *Child's Nervous System* 37, no. 9 (2021): 2723–24, https://doi.org/10.1007/s00381-020-04564-z.

47 **Dr. Adrian Upton:** Boyce Rensberger, "Jell-O Test Finds Lifelike Signal," *New York Times*, March 6, 1976, sec. Archives, https://www.nytimes.com/1976/03/06/archives/jello-test-finds-lifelike-signal-blob-gives-confusing-signs-in.html.

47 **Tragically, Berger died:** Karthik Meda, "Hans Berger and the Electroencephalogram (P1-1.016)," *Neurology* 102, no. 17, suppl. 1 (2024): 5587, https://doi.org/10.1212/WNL.0000000000205921.

48 **With a lifetime prevalence:** Helen M. Stallman and Mark Kohler, "Prevalence of Sleepwalking: A Systematic Review and Meta-Analysis," *PLoS One* 11, no. 11 (2016): e0164769, https://doi.org/10.1371/journal.pone.0164769.

48 **identical twins are more likely:** C. Hublin, J. Kaprio, M. Partinen, K. Heikkilä, and M. Koskenvuo, "Prevalence and Genetics of Sleepwalking: A Population-Based Twin Study," *Neurology* 48, no. 1 (1997): 177–81, https://doi.org/10.1212/wnl.48.1.177.

49 **Medieval records of sleepwalkers:** William MacLehose, "Sleepwalking, Violence and Desire in the Middle Ages," *Culture, Medicine and Psychiatry* 37, no. 4 (2013): 601–24, https://doi.org/10.1007/s11013-013-9344-9.

49 **In the case of Lee Hadwin:** "Meet Lee Hadwin the 'Sleepwalking Artist' Who Can't Draw When He's Awake," *The Independent*, January 14, 2015, https://www.independent.co.uk/arts-entertainment/art/features/forget-tracey-emin-s-bed-meet-the-sleep-walking-artist-9977382.html.

49 **A three-year-old girl:** Katherine R. Lichtblau, Patrick W. Romani, Brian D. Greer, Wayne W. Fisher, and Allie K. Bragdon, "Remote Treatment of Sleep-Related Trichotillomania and Trichophagia," *Journal of Applied Behavior Analysis* 51, no. 2 (2018): 255–62, https://doi.org/10.1002/jaba.442.

49 **A 37-year-old man:** Ranji Varghese, Jorge Rey de Castro, Cesar Liendo, and Carlos H. Schenck, "Two Cases of Sleep-Related Eating Disorder Responding Promptly to Low-Dose Sertraline Therapy," *Journal of Clinical Sleep Medicine* 14, no. 10 (2018): 1805–8, https://doi.org/10.5664/jcsm.7404.

50 **A 45-year-old woman:** Sana Elham Kazi, Jamal Mujaddid M. Mohammed, and Carlos H. Schenck, "Sleepwalking, Sleep-Related Eating Disorder and Sleep-Related Smoking Successfully Treated with Topiramate: A Case Report." *Sleep Science* 15, no. 3 (2022): 370–73, https://doi.org/10.5935/1984-0063.20220065.

50 **And there's the very surprising case of Kenneth James Parks:** Roger Broughton, Rodger Billings, Rosalind Cartwright, David Doucette, J. Edmeads, Marlys Edwardh, F. Ervin, Beverley Orchard, R. Hill, and Gavin Turrell, "Homicidal Somnambulism: A Case Report," *Sleep* 17, no. 3 (1994): 253–64.

51 **he was acquitted on both charges:** "R. v. Parks, 1992 CanLII 78 (SCC), [1992] 2 SCR 871," https://www.canlii.org/en/ca/scc/doc/1992/1992canlii78/1992canlii78.html.

51 **Isom Bradley was charged:** Beth E. Teacher, "Sleepwalking Used as a Defense in Criminal Cases and the Evolution of the Ambien Defense," *Duquesne Criminal Law Journal* 2, no. 2 (2016): 1, https://dsc.duq.edu/dclj/vol2/iss2/6.

52 **spouses needing dental repairs:** Eric J. Olson, Bradley F. Boeve, and Michael H. Silber, "Rapid Eye Movement Sleep Behaviour Disorder: Demographic, Clinical and Laboratory Findings in 93 Cases," *Brain* 123, no. 2 (2000): 331–39, https://doi .org/10.1093/brain/123.2.331.

52 **spouses being bit:** Ana Fernández-Arcos, Alex Iranzo, Mónica Serradell, Carles Gaig, and Joan Santamaria, "The Clinical Phenotype of Idiopathic Rapid Eye Movement Sleep Behavior Disorder at Presentation: A Study in 203 Consecutive Patients," *Sleep* 39, no. 1 (2016): 121–32, https://doi.org/10.5665/sleep.5332.

52 **Acting out a football dream . . . a dream-bear attack:** Stuart J. McCarter, Erik K. St. Louis, Christopher L. Boswell, Lucas G. Dueffert, Nancy Slocumb, Bradley F. Boeve, Michael H. Silber, Eric J. Olson, Timothy I. Morgenthaler, and Maja Tippmann-Peikert, "Factors Associated with Injury in REM Sleep Behavior Disorder," *Sleep Medicine* 15, no. 11 (2014): 1332–38, https://doi.org/10.1016/j .sleep.2014.06.002.

52 **A woman dreamed of being chased:** Zhen Fan, Yanrui Niu, and Hui Zhang, "Case Report of Rapid-Eye-Movement (REM) Sleep Behavior Disorder," *Shanghai Archives of Psychiatry* 25, no. 2 (2013): 121–23, https://doi.org/10.3969/j.issn.1002 -0829.2013.02.010.

53 **could be the first suggestion that the trait runs in families:** Eric Lewin Altschuler, "Hereditary Somnambulism in Dracula," *Journal of the Royal Society of Medicine* 96, no. 1 (2003): 51–52, https://doi.org/10.1177/014107680309600124.

53 **racked with guilt:** "Sleepwalking and Guilt," *British Medical Journal* 2, no. 5702 (1970): 186, https://www.ncbi.nlm.nih.gov/pmc/articles/PMC1699981/?page=2.

54 **These motor inhibition systems are most active:** Alex Iranzo, "The REM Sleep Circuit and How Its Impairment Leads to REM Sleep Behavior Disorder," *Cell and Tissue Research* 373, no. 1 (2018): 245–66, https://doi.org/10.1007/s00441-018 -2852-8.

54 **One part of the brain stem:** J. C. Hendricks, A. R. Morrison, and G. L. Mann, "Different Behaviors during Paradoxical Sleep without Atonia Depend on Pontine Lesion Site," *Brain Research* 239, no. 1 (1982): 81–105, https://doi .org/10.1016/0006-8993(82)90835-6.

55 **Prescription insomnia drugs:** Helen M. Stallman, Mark Kohler, and Jason White, "Medication Induced Sleepwalking: A Systematic Review," *Sleep Medicine Reviews* 37 (February 2018): 105–13, https://doi.org/10.1016/j.smrv.2017.01.005.

55 **spilled gasoline:** Barry Liskow and Andrei Pikalov, "Zaleplon Overdose Associated with Sleepwalking and Complex Behavior," *Journal of the American Academy of Child and Adolescent Psychiatry* 43, no. 8 (2004): 927–28, https://doi.org/10.1097/01 .chi.0000129219.66563.aa.

55 **a man who shot himself:** Christopher E. Gibson and Jason P. Caplan, "Zolpidem-Associated Parasomnia with Serious Self-Injury: A Shot in the Dark," *Psychosomatics* 52, no. 1 (2011): 88–91, https://doi.org/10.1016/j.psym.2010.11.015.

55 **About one-fifth:** Brian A. Sharpless and Jacques P. Barber, "Lifetime Prevalence Rates of Sleep Paralysis: A Systematic Review," *Sleep Medicine Reviews* 15, no. 5 (2011): 311–15, https://doi.org/10.1016/j.smrv.2011.01.007.

56 **Some link alien abduction:** Richard J. McNally and Susan A. Clancy, "Sleep Paralysis, Sexual Abuse, and Space Alien Abduction," *Transcultural Psychiatry* 42, no. 1 (2005): 113–22, https://doi.org/10.1177/1363461505050715.

56 **Egyptians attributed:** Baland Jalal, Joseph Simons-Rudolph, Bamo Jalal, and Devon E. Hinton, "Explanations of Sleep Paralysis among Egyptian College Students and the General Population in Egypt and Denmark," *Transcultural Psychiatry* 51, no. 2 (2014): 158–75, https://doi.org/10.1177/1363461513503378.

56 **Some Italians:** Baland Jalal, Andrea Romanelli, and Devon E. Hinton, "Cultural Explanations of Sleep Paralysis in Italy: The Pandafeche Attack and Associated Supernatural Beliefs," *Culture, Medicine and Psychiatry* 39, no. 4 (2015): 651–64, https://doi.org/10.1007/s11013-015-9442-y.

56 **Some Cambodian refugees:** Devon E. Hinton, Vuth Pich, Dara Chhean, and Mark H. Pollack, "'The Ghost Pushes You Down': Sleep Paralysis-Type Panic Attacks in a Khmer Refugee Population," *Transcultural Psychiatry* 42, no. 1 (2005): 46–77, https://doi.org/10.1177/1363461505050710.

57 **The condition is also called Ondyne's curse:** Zeferino Demartini, Luana Antunes Maranha Gatto, Gelson Luis Koppe, Alexandre Novicki Francisco, and Enio Eduardo Guerios, "Ondine's Curse: Myth Meets Reality," *Sleep Medicine: X* 2 (December 2020): 100012, https://doi.org/10.1016/j.sleepx.2020.100012.

57 **A neurochemical signal:** S. Nishino, B. Ripley, S. Overeem, G. J. Lammers, and E. Mignot, "Hypocretin (Orexin) Deficiency in Human Narcolepsy," *Lancet* 355, no. 9197 (2000): 39–40, https://doi.org/10.1016/S0140-6736(99)05582-8.

58 **had less than one-tenth:** T. C. Thannickal, R. Y. Moore, R. Nienhuis, L. Ramanathan, S. Gulyani, M. Aldrich, M. Cornford, and J. M. Siegel, "Reduced Number of Hypocretin Neurons in Human Narcolepsy," *Neuron* 27, no. 3 (2000): 469–74, https://doi.org/10.1016/s0896-6273(00)00058-1.

58 **development of psychology:** D. Westen, "The Scientific Legacy of Sigmund Freud: Toward a Psychodynamically Informed Psychological Science," *Psychological Bulletin* 124. no. 3 (1988): 333–71, https://doi.org/10.1037/0033-2909.124.3.333.

58 **Today's forms of talk therapy:** Alan F. Javel, "The Freudian Antecedents of Cognitive-Behavioral Therapy," *Journal of Psychotherapy Integration* 9, no. 4 (1999): 397–407, https://doi.org/10.1023/A:1023247428670.

59 **the Y chromosome contributes:** J. M. Tanner, A. Prader, H. Habich, and M. A. Ferguson-Smith, "Genes on the Y Chromosome Influencing Rate of Maturation in Man: Skeletal Age Studies in Children with Klinefelter's (XXY) and Turner's (XO) Syndromes," *Lancet* 274, no. 7095 (1959): 141–44.

60 **Freud provides a framework:** "A Science Odyssey: People and Discoveries: Freud's Book, 'The Interpretation of Dreams' Released," n.d., https://www.pbs.org/wgbh/aso/databank/entries/dh00fr.html.

61 **dreams help us reinforce spatial memories:** Claudia Picard-Deland, Giulio Bernardi, Lisa Genzel, Martin Dresler, and Sarah F. Schoch, "Memory Reactivations

during Sleep: A Neural Basis of Dream Experiences?," *Trends in Cognitive Sciences* 27, no. 6 (2023): 568–82, https://doi.org/10.1016/j.tics.2023.02.006.

62 **Dreams often have some spatial content . . . unlearning takes place:** Robert P. Vertes, "Memory Consolidation in Sleep; Dream or Reality," *Neuron* 44, no. 1 (2004): 135–48, https://doi.org/10.1016/j.neuron.2004.08.034.

63 **Blind people can learn to echolocate:** Liam J. Norman and Lore Thaler, "Retinotopic-Like Maps of Spatial Sound in Primary 'Visual' Cortex of Blind Human Echolocators," *Proceedings of the Royal Society B: Biological Sciences* 286, no. 1912 (2019): 20191910, https://doi.org/10.1098/rspb.2019.1910.

63 **when blind people read Braille:** N. Sadato and M. Hallett, "fMRI Occipital Activation by Tactile Stimulation in a Blind Man," *Neurology* 52, no. 2 (1999): 423, https://doi.org/10.1212/wnl.52.2.423.

63 **One fascinating proposition:** David M. Eagleman and Don A. Vaughn, "The Defensive Activation Theory: REM Sleep as a Mechanism to Prevent Takeover of the Visual Cortex," *Frontiers in Neuroscience* 15 (May 2021): 632853, https://doi.org/10.3389/fnins.2021.632853.

64 **help us flex our creativity:** Erik Hoel, "The Overfitted Brain: Dreams Evolved to Assist Generalization," *Patterns* 2, no. 5 (2021): 100244, https://doi.org/10.1016/j.patter.2021.100244.

CHAPTER 4

66 **1,000 pills a week:** "Philip K. Dick: A Pulp Sci-Fi Writer Finally Wins Respect," *New York Times*, May 9, 2007, sec. Arts, https://www.nytimes.com/2007/05/09/arts/09iht-pulp.1.5634129.html.

66 **68 pages of prose:** "Interview with Philip K. Dick," May 11, 2012, https://web.archive.org/web/20120511085758/http://philipkdick.com/media_sfeye96.html.

68 **he taught his pigeons to play competitive table tennis:** Marina Koren, "B. F. Skinner: The Man Who Taught Pigeons to Play Ping-Pong and Rats to Pull Levers," March 20, 2013, https://www.smithsonianmag.com/science-nature/bf-skinner-the-man-who-taught-pigeons-to-play-ping-pong-and-rats-to-pull-levers-5363946.

68 **government-funded Project Pigeon:** Benjamin Schultz-Figueroa, "Project Pigeon: Rendering the War Animal through Optical Technology," in *The Celluloid Specimen: Moving Image Research into Animal Life* (Berkeley: University of California Press, 2023), 137–54, http://www.jstor.org/stable/jj.1791902.16.

69 **a few improvements over the years:** Sandeep Sharma, Cecile Hryhorczuk, and Stephanie Fulton, "Progressive-Ratio Responding for Palatable High-Fat and High-Sugar Food in Mice," *Journal of Visualized Experiments*, no. 63 (May 2012): e3754, https://doi.org/10.3791/3754.

70 **gain access to other mice:** Leslie A. Ramsey, Fernanda M. Holloman, Samantha S. Lee, and Marco Venniro, "An Operant Social Self-Administration and Choice Model in Mice," *Nature Protocols* 18, no. 6 (2023): 1669–86, https://doi.org/10.1038/s41596-023-00813-y.

70 **will learn to press to avoid pain:** Isaac R. Galatzer-Levy, Justin Moscarello, Esther M. Blessing, JoAnna Klein, Christopher K. Cain, and Joseph E. LeDoux,

"Heterogeneity in Signaled Active Avoidance Learning: Substantive and Methodological Relevance of Diversity in Instrumental Defensive Responses to Threat Cues," *Frontiers in Systems Neuroscience* 8 (September 2014): 179, https://doi.org/10.3389/fnsys.2014.00179.

71 **many drugs that humans take:** Eoin C. O'Connor, Kathryn Chapman, Paul Butler, and Andy N. Mead, "The Predictive Validity of the Rat Self-Administration Model for Abuse Liability," *Neuroscience and Biobehavioral Reviews* 35, no. 3 (2011): 912–38, https://doi.org/10.1016/j.neubiorev.2010.10.012.

71 **The final end point:** G. Deneau, T. Yanagita, and M. H. Seevers, "Self-Administration of Psychoactive Substances by the Monkey," *Psychopharmacologia* 16, no. 1 (1969): 30–48, https://doi.org/10.1007/BF00405254.

71 **humans are diagnosed with substance use disorder:** Deborah S. Hasin, Charles P. O'Brien, Marc Auriacombe, Guilherme Borges, Kathleen Bucholz, Alan Budney, Wilson M. Compton, Thomas Crowley, Walter Ling, Nancy M. Petry, Marc Schuckit, and Bridget F. Grant, "DSM-5 Criteria for Substance Use Disorders: Recommendations and Rationale," *American Journal of Psychiatry* 170, no. 8 (2013): 834–51, https://doi.org/10.1176/appi.ajp.2013.12060782.

72 **drive hundreds of miles:** Brakkton Booker, "Doctor Gets 40 Years for Illegally Prescribing More Than Half a Million Opioid Doses," October 2, 2019, https://www.npr.org/2019/10/02/766403612/doctor-gets-40-years-for-illegally-prescribing-more-than-half-a-million-opioid-d.

73 **the influence of environment on drug-taking behavior:** Patricia F. Hadaway, Bruce K. Alexander, Robert B. Coambs, and Barry Beyerstein, "The Effect of Housing and Gender on Preference for Morphine-Sucrose Solutions in Rats," *Psychopharmacology* 66, no. 1 (1979): 87–91, https://doi.org/10.1007/BF00431995.

73 **Debi Austin of 1990s:** "The Debi Austin Story," California Tobacco Control Program, February 23, 2022, https://www.undo.org/exposed/the-debi-austin-story.

73 **Others persist:** Yann Pelloux, Barry J. Everitt, and Anthony Dickinson, "Compulsive Drug Seeking by Rats under Punishment: Effects of Drug Taking History." *Psychopharmacology* 194, no. 1 (2007): 127–37, https://doi.org/10.1007/s00213-007-0805-0.

74 **Rats with a history of drug exposure:** Susan Schenk and Brian Partridge, "Sensitization and Tolerance in Psychostimulant Self-Administration," *Pharmacology Biochemistry and Behavior* 57, no. 3 (1997): 543–50, https://doi.org/10.1016/S0091-3057(96)00447-9.

74 **an opioid-dependent rat:** Kelly E. Dunn, Andrew S. Huhn, Cecilia L. Bergeria, Cassandra D. Gipson, and Elise M. Weerts, "Non-Opioid Neurotransmitter Systems That Contribute to the Opioid Withdrawal Syndrome: A Review of Preclinical and Human Evidence," *Journal of Pharmacology and Experimental Therapeutics* 371, no. 2 (2019): 422–52, https://doi.org/10.1124/jpet.119.258004.

75 **natural redheads are more sensitive:** Kathleen C. Robinson, Lajos V. Kemény, Gillian L. Fell, Andrea L. Hermann, Jennifer Allouche, Weihua Ding, Ajay Yekkirala, Jennifer J. Hsiao, Mack Y. Su, Nicholas Theodosakis, Gabor Kozak, Yuichi Takeuchi, Shiqian Shen, Antal Berenyi, Jianren Mao, Clifford J. Woolf, and David E. Fisher, "Reduced

MC4R Signaling Alters Nociceptive Thresholds Associated with Red Hair," *Science Advances* 7, no. 14 (2021): eabd1310, https://doi.org/10.1126/sciadv.abd1310.

75 **Adolescent rats behave:** Marci R. Mitchell, Virginia G. Weiss, B. Sofia Beas, Drake Morgan, Jennifer L. Bizon, and Barry Setlow, "Adolescent Risk Taking, Cocaine Self-Administration, and Striatal Dopamine Signaling," *Neuropsychopharmacology* 39, no. 4 (2014): 955–62, https://doi.org/10.1038/npp.2013.295.

75 **tacking a proton-hungry fluorine:** V. Spahn, G. Del Vecchio, D. Labuz, A. Rodriguez-Gaztelumendi, N. Massaly, J. Temp, V. Durmaz, P. Sabri, M. Reidelbach, H. Machelska, M. Weber, and C. Stein, "A Nontoxic Pain Killer Designed by Modeling of Pathological Receptor Conformations," *Science* 355, no. 6328 (2017): 966–69, https://doi.org/10.1126/science.aai8636.

75 **inject an agent:** Michael D. Raleigh, Paul R. Pentel, and Mark G. LeSage, "Pharmacokinetic Correlates of the Effects of a Heroin Vaccine on Heroin Self-Administration in Rats," *PLoS One* 9, no. 12 (2014): e115696, https://doi.org/10.1371/journal.pone.0115696.

75 **monkeys with this immune treatment:** K. F. Bonese, B. H. Wainer, F. W. Fitch, R. M. Rothberg, and C. R. Schuster, "Changes in Heroin Self-Administration by a Rhesus Monkey after Morphine Immunisation," *Nature* 252, no. 5485 (1974): 708–10, https://doi.org/10.1038/252708a0.

76 **The outlier of the book is chapter 9:** Wikisource contributors, "Drug Themes in Science Fiction," https://en.wikisource.org/w/index.php?title=Drug_Themes_in_Science_Fiction&oldid=3710166 (accessed July 16, 2024).

77 **In a Skinner box:** C. David Wise and Larry Stein, "Facilitation of Brain Self-Stimulation by Central Administration of Norepinephrine," *Science* 163, no. 3864 (1969): 299–301, https://doi.org/10.1126/science.163.3864.299.

77 **a tiny four-paragraph article:** Arvid Carlsson, Margit Lindqvist, Tor Magnusson, and Bertil Waldeck, "On the Presence of 3-Hydroxytyramine in Brain," *Science* 127, no. 3296 (1958): 471, https://doi.org/10.1126/science.127.3296.471.a.

77 **Wise showed:** Roy A. Wise, "Dopamine and Reward: The Anhedonia Hypothesis 30 Years On," *Neurotoxicity Research* 14, nos. 2–3 (2008): 169–83, https://doi.org/10.1007/BF03033808.

78 **pharmaceutical companies would routinely measure:** David J. Nutt, Anne Lingford-Hughes, David Erritzoe, and Paul R. A. Stokes, "The Dopamine Theory of Addiction: 40 Years of Highs and Lows," *Nature Reviews Neuroscience* 16, no. 5 (2015): 305–12, https://doi.org/10.1038/nrn3939.

78 **Instead of the occasional:** Jacques Mirenowicz and Wolfram Schultz, "Preferential Activation of Midbrain Dopamine Neurons by Appetitive Rather Than Aversive Stimuli," *Nature* 379, no. 6564 (1996): 449–51, https://doi.org/10.1038/379449a0.

79 **According to the reward-prediction error hypothesis:** Wolfram Schultz, "Predictive Reward Signal of Dopamine Neurons," *Journal of Neurophysiology* 80, no. 1 (1998): 1–27, https://doi.org/10.1152/jn.1998.80.1.1.

79 **Most recreational drugs interfere:** Eric J. Nestler and Robert C. Malenka, "The Addicted Brain," *Scientific American* 290, no. 3 (2004): 78–85, https://doi.org/10.1038/scientificamerican0304-78.

81 **If you give ergot:** Thomas Haarmann, Yvonne Rolke, Sabine Giesbert, and Paul Tudzynski, "Ergot: From Witchcraft to Biotechnology," *Molecular Plant Pathology* 10, no. 4 (2009): 563–77, https://doi.org/10.1111/j.1364-3703.2009.00548.x.

82 **When the trials began:** Linnda R. Caporael, "Ergotism: The Satan Loosed in Salem?," *Science* 192, no. 4234 (1976): 21–26, https://doi.org/10.1126/science.769159.

82 **Art historians suggest:** Fiona Macdonald, "Hidden Meanings in the Garden of Earthly Delights," August 9, 2016, https://www.bbc.com/culture/article/20160809-hidden-meanings-in-the-garden-of-earthly-delights.

83 **In 1938, a Swiss chemist:** Albert Hofmann, *LSD, My Problem Child: Reflections on Sacred Drugs, Mysticism and Science* (Santa Cruz, CA: Multidisciplinary Association for Psychedelic Studies, 2009).

85 **Small trials have demonstrated applications for treating:** Robin L. Carhart-Harris and Guy M. Goodwin, "The Therapeutic Potential of Psychedelic Drugs: Past, Present, and Future," *Neuropsychopharmacology* 42, no. 11 (2017): 2105–13, https://doi.org/10.1038/npp.2017.84.

86 **following in the footsteps of Denver:** "In Close Vote, Denver Becomes 1st U.S. City to Decriminalize Psychedelic Mushrooms," n.d., https://www.npr.org/sections/health-shots/2019/05/09/721660053/in-close-vote-denver-becomes-first-u-s-city-to-decriminalize-psychedelic-mushroo.

86 **Much work has yet to be done:** Liam Drew, "Your Brain on Psychedelics," *Nature* 609, no. 7929 (2022): S92–94, https://doi.org/10.1038/d41586-022-02874-7.

CHAPTER 5

89 **Alcohol also modifies:** M. Trevisani, D. Smart, M. J. Gunthorpe, M. Tognetto, M. Barbieri, B. Campi, S. Amadesi, J. Gray, J. C. Jerman, S. J. Brough, D. Owen, G. D. Smith, A. D. Randall, S. Harrison, A. Bianchi, J. B. Davis, and P. Geppetti, "Ethanol Elicits and Potentiates Nociceptor Responses via the Vanilloid Receptor-1." *Nature Neuroscience* 5, no. 6 (2002): 546–51, https://doi.org/10.1038/nn0602-852.

89 **Piezo1 and Piezo2 receptors:** Scott L. Earley, Fernando Santana, and W. Jonathan Lederer, "The Physiological Sensor Channels TRP and Piezo: Nobel Prize in Physiology or Medicine 2021," *Physiological Reviews* 102, no. 2 (2022): 1153–58, https://doi.org/10.1152/physrev.00057.2021.

90 **TRPA1 receptors:** Daniel Souza Monteiro de Araujo, Romina Nassini, Pierangelo Geppetti, and Francesco De Logu, "TRPA1 as a Therapeutic Target for Nociceptive Pain," *Expert Opinion on Therapeutic Targets* 24, no. 10 (2020): 997–1008, https://doi.org/10.1080/14728222.2020.1815191.

90 **written by the late Justin Schmidt:** Justin O. Schmidt, *The Sting of the Wild* (Baltimore: Johns Hopkins University Press, 2018).

91 **A 34-year-old man breaks his hand:** "David Haye Has Achilles Surgery after Tony Bellew Defeat," *BBC Sport*, March 5, 2017, sec. Boxing, https://www.bbc.com/sport/boxing/39171240.

92 **In a reenactment of the Passion of Jesus Christ:** "Surviving Crucifixion: An Interview with Ruben Enaje" (editorial), *The Review of Religions*, April 2, 2021,

https://www.reviewofreligions.org/29953/surviving-crucifixion-an-interview-with-ruben-enaje.

93 **In preparation for the ritual:** Anand N. Bosmia, Christoph J. Griessenauer, Vidal Haddad, and R. Shane Tubbs, "Ritualistic Envenomation by Bullet Ants among the Sateré-Mawé Indians in the Brazilian Amazon," *Wilderness and Environmental Medicine* 26, no. 2 (2015): 271–73, https://doi.org/10.1016/j.wem.2014.09.003.

93 **Human imaging studies have identified:** J. D. Talbot, S. Marrett, A. C. Evans, E. Meyer, M. C. Bushnell, and G. H. Duncan, "Multiple Representations of Pain in Human Cerebral Cortex," *Science* 251, no. 4999 (1991): 1355–58, https://doi.org/10.1126/science.2003220.

93 **anterior cingulate cortex:** Tor D. Wager, Lauren Y. Atlas, Martin A. Lindquist, Mathieu Roy, Choong-Wan Woo, and Ethan Kross, "An fMRI-Based Neurologic Signature of Physical Pain," *New England Journal of Medicine* 368, no. 15 (2013): 1388–97, https://doi.org/10.1056/NEJMoa1204471.

94 **To model socioemotional pain:** Naomi I. Eisenberger, Matthew D. Lieberman, and Kipling D. Williams, "Does Rejection Hurt? An FMRI Study of Social Exclusion," *Science* 302, no. 5643 (2003): 290–92, https://doi.org/10.1126/science.1089134.

95 **This idea, gate control theory:** R. Melzack and P. D. Wall, "Pain Mechanisms: A New Theory," *Science* 150, no. 3699 (1965): 971–79, https://doi.org/10.1126/science.150.3699.971.

95 **One man, however, might:** Otto Muzik, Kaice T. Reilly, and Vaibhav A. Diwadkar, "'Brain over Body'—A Study on the Willful Regulation of Autonomic Function during Cold Exposure," *NeuroImage* 172 (May 2018): 632–41, https://doi.org/10.1016/j.neuroimage.2018.01.067.

96 **the more your attention is drawn away:** Katharina M. Rischer, Ana M. González-Roldán, Pedro Montoya, Sandra Gigl, Fernand Anton, and Marian van der Meulen, "Distraction from Pain: The Role of Selective Attention and Pain Catastrophizing," *European Journal of Pain* 24, no. 10 (2020): 1880–91, https://doi.org/10.1002/ejp.1634.

96 **In another task called the Stroop task:** Susanna J. Bantick, Richard G. Wise, Alexander Ploghaus, Stuart Clare, Stephen M. Smith, and Irene Tracey, "Imaging How Attention Modulates Pain in Humans Using Functional MRI," *Brain* 125, no. 2 (2002): 310–19, https://doi.org/10.1093/brain/awf022.

96 **But my favorite painkiller:** Richard Stephens and Olly Robertson, "Swearing as a Response to Pain: Assessing Hypoalgesic Effects of Novel 'Swear' Words," *Frontiers in Psychology* 11 (April 2020): 723, https://doi.org/10.3389/fpsyg.2020.00723.

97 **Unsurprisingly, pain is the principal complaint . . . chronic pain affects up to 30 percent:** Steven P. Cohen, Lene Vase, and William M. Hooten, "Chronic Pain: An Update on Burden, Best Practices, and New Advances," *Lancet* 397, no. 10289 (2021): 2082–97, https://doi.org/10.1016/S0140-6736(21)00393-7.

97 **Local inflammation leads to hypersensitivity:** Marwan N. Baliki and A. Vania Apkarian, "Nociception, Pain, Negative Moods, and Behavior Selection," *Neuron* 87, no. 3 (2015): 474–91, https://doi.org/10.1016/j.neuron.2015.06.005.

98 **Consider a study in which rats learned:** R. G. Phillips and J. E. LeDoux, "Differential Contribution of Amygdala and Hippocampus to Cued and Contextual Fear Conditioning," *Behavioral Neuroscience* 106, no. 2 (1992): 274–85, https://doi.org/10.1037//0735-7044.106.2.274.

99 **According to self-report scores:** Donald A. Redelmeier, Joel Katz, and Daniel Kahneman, "Memories of Colonoscopy: A Randomized Trial," *Pain* 104, no. 1 (2003): 187–94, https://journals.lww.com/pain/fulltext/2003/07000/memories_of_colonoscopy__a_randomized_trial.20.aspx.

101 **Up to 85 percent of amputees:** Jacob Boomgaardt, Kovosh Dastan, Tiffany Chan, Ashley Shilling, Alaa Abd-Elsayed, and Lynn Kohan, "An Algorithm Approach to Phantom Limb Pain," *Journal of Pain Research* 15 (October 2022): 3349–67, https://doi.org/10.2147/JPR.S355278.

101 **phantom erection pain:** Popy Arizona, Erikavitri Yulianti, and Izzatul Fithriyah, "Psychiatric Approach in Phantom Erection Postpenectomy Patient," *Case Reports in Psychiatry* 2023: 4113455, https://doi.org/10.1155/2023/4113455.

101 **Internal organs are also:** T. L. Dorpat, "Phantom Sensations of Internal Organs," *Comprehensive Psychiatry* 12, no. 1 (1971): 27–35, https://doi.org/10.1016/0010-440X(71)90053-8.

102 **dunk in boiling oil:** Philippe Hernigou, "Ambroise Paré II: Paré's Contributions to Amputation and Ligature," *International Orthopaedics* 37, no. 4 (2013): 769–72, https://doi.org/10.1007/s00264-013-1857-x.

102 **I tried to get at it to rub it:** S. Weir Mitchell *The Autobiography of a Quack, and the Case of George Dedlow* (Salt Lake City, UT: Project Gutenberg, 2006), https://www.gutenberg.org/ebooks/693.

102 **Mitchell chose *Atlantic Monthly*:** Annie Woodhouse, "Phantom Limb Sensation," *Clinical and Experimental Pharmacology and Physiology* 32, nos. 1–2 (2005): 132–34, https://doi.org/10.1111/j.1440-1681.2005.04142.x.

103 **a problem with the peripheral nervous system:** Bishnu Subedi and George T. Grossberg, "Phantom Limb Pain: Mechanisms and Treatment Approaches," *Pain Research and Treatment* 2011: 864605, https://doi.org/10.1155/2011/864605.

103 **people born without limbs can experience phantom pains:** Martin Diers, Xaver Fuchs, Robin Bekrater-Bodmann, and Herta Flor, "Prevalence of Phantom Phenomena in Congenital and Early-Life Amputees," *Journal of Pain* 24, no. 3 (2023): 502–8, https://doi.org/10.1016/j.jpain.2022.10.010.

104 **scratching the face can help:** Katja Guenther, "'It's All Done with Mirrors': V. S. Ramachandran and the Material Culture of Phantom Limb Research," *Medical History* 60, no. 3 (2016): 342–58, https://doi.org/10.1017/mdh.2016.27.

105 **jumping off a roof:** Richard Knox, "Newly Found Gene Mutation Banishes Pain," December 13, 2006, https://www.npr.org/2006/12/13/6620733/newly-found-gene-mutation-banishes-pain.

105 **pain insensitivity exists on a spectrum:** Michaela Auer-Grumbach, "Chapter 50—Hereditary Sensory and Autonomic Neuropathies," *Handbook of Clinical Neurology* 115 (2013): 893–906, https://doi.org/10.1016/B978-0-444-52902-2.00050-3.

106 **two brothers:** A. G. Swanson, G. C. Buchan, and E. C. Alvord, "Anatomic Changes in Congenital Insensitivity to Pain: Absence of Small Primary Sensory Neurons in Ganglia, Roots, and Lissauer's Tract," *Archives of Neurology* 12, no. 1 (1965): 12–18, https://doi.org/10.1001/archneur.1965.00460250016002.

106 **of schizophrenia-induced pain insensitivity:** Akhil Kallur, Eungjae Yoo, Fred Bien-Aime, and Hussam Ammar, "Diagnostic Overshadowing and Pain Insensitivity in a Schizophrenic Patient with Perforated Duodenal Ulcer," *Cureus* 14, no. 2 (2022): e21800, https://doi.org/10.7759/cureus.21800.

106 **heart attacks without any pain:** W. E. Marchand, "Occurrence of Painless Myocardial Infarction in Psychotic Patients," *New England Journal of Medicine* 253, no. 2 (1955): 51–55, https://doi.org/10.1056/NEJM195507142530202.

106 **a mutation of SCN9A:** James J. Cox, Frank Reimann, Adeline K. Nicholas, Gemma Thornton, Emma Roberts, Kelly Springell, Gulshan Karbani, Hussain Jafri, Jovaria Mannan, Yasmin Raashid, Lihadh Al-Gazali, Henan Hamamy, Enza Maria Valente, Shaun Gorman, Richard Williams, Duncan P. McHale, John N. Wood, Fiona M. Gribble, and C. Geoffrey Woods, "An SCN9A Channelopathy Causes Congenital Inability to Experience Pain," *Nature* 444, no. 7121 (2006): 894–98, https://doi.org/10.1038/nature05413.

106 **found on olfactory neurons:** Ichrak Drissi, William Aidan Woods, and Christopher Geoffrey Woods, "Understanding the Genetic Basis of Congenital Insensitivity to Pain," *British Medical Bulletin* 133, no. 1 (2020): 65–78, https://doi.org/10.1093/bmb/ldaa003.

106 **called pain asymbolia:** Colin Klein, "What Pain Asymbolia Really Shows," *Mind* 124, no. 494 (2015): 493–516, https://doi.org/10.1093/mind/fzu185.

107 **Brain-imaging scans of asymbolics:** M. Berthier, S. Starkstein, and R. Leiguarda, "Asymbolia for Pain: A Sensory-Limbic Disconnection Syndrome," *Annals of Neurology* 24, no. 1 (1988): 41–49, https://doi.org/10.1002/ana.410240109.

109 **cocaine on the gums was Sigmund Freud:** A. López-Valverde, J. de Vicente, L. Martínez-Domínguez, and R. Gómez de Diego, "Local Anaesthesia through the Action of Cocaine, the Oral Mucosa and the Vienna Group," *British Dental Journal* 217, no. 1 (2014): 41–43, https://doi.org/10.1038/sj.bdj.2014.546.

109 **Since opioids also disinhibit:** Barbara Juarez and Ming-Hu Han, "Diversity of Dopaminergic Neural Circuits in Response to Drug Exposure," *Neuropsychopharmacology* 41, no. 10 (2016): 2424–46, https://doi.org/10.1038/npp.2016.32.

110 **densely in the anterior cingulate:** B. A. Vogt, R. G. Wiley, and E. L. Jensen, "Localization of Mu and Delta Opioid Receptors to Anterior Cingulate Afferents and Projection Neurons and Input/Output Model of Mu Regulation," *Experimental Neurology* 135, no. 2 (1995): 83–92, https://doi.org/10.1006/exnr.1995.1069.

110 **Antagonists such as naloxone:** Michael S. Minett, Vanessa Pereira, Shafaq Sikandar, Ayako Matsuyama, Stéphane Lolignier, Alexandros H. Kanellopoulos, Flavia Mancini, Gian D. Iannetti, Yury D. Bogdanov, Sonia Santana-Varela, Queensta Millet, Giorgios Baskozos, Raymond MacAllister, James J. Cox, Jing Zhao, and John N. Wood, "Endogenous Opioids Contribute to Insensitivity to Pain in

Humans and Mice Lacking Sodium Channel Na_v1.7," *Nature Communications* 6 (December 2015): 8967, https://doi.org/10.1038/ncomms9967.

CHAPTER 6

113 **A group of chess players:** F. Gobet, P. C. R. Lane, S. Croker, P. C.-H. Cheng, G. Jones, I. Oliver, and J. M. Pine, "Chunking Mechanisms in Human Learning," *Trends in Cognitive Sciences* 5, no. 6 (2001): 236–43, https://doi.org/10.1016/s1364-6613(00)01662-4.

114 **is the engram:** Sheena A. Josselyn and Susumu Tonegawa, "Memory Engrams: Recalling the Past and Imagining the Future," *Science* 367, no. 6473 (2020): eaaw4325, https://doi.org/10.1126/science.aaw4325.

115 **plasticity might look:** Cassiano Ricardo Alves Faria Diniz and Ana Paula Crestani, "The Times They Are A-Changin': A Proposal on How Brain Flexibility Goes beyond the Obvious to Include the Concepts of 'Upward' and 'Downward' to Neuroplasticity," *Molecular Psychiatry* 28, no. 3 (2023): 977–92, https://doi.org/10.1038/s41380-022-01931-x.

116 **The experiments demonstrating learning in *Aplysia*:** Eric R. Kandel, *In Search of Memory: The Emergence of a New Science of Mind* (New York: Norton, 2006).

117 **the hippocampus of anesthetized rabbits:** T. V. Bliss and T. Lomo, "Long-Lasting Potentiation of Synaptic Transmission in the Dentate Area of the Anaesthetized Rabbit Following Stimulation of the Perforant Path," *Journal of Physiology* 232, no. 2 (1973): 331–56, https://doi.org/10.1113/jphysiol.1973.sp010273.

117 **can persist for up to a year:** Wickliffe C. Abraham, Barbara Logan, Jeffrey M. Greenwood, and Michael Dragunow, "Induction and Experience-Dependent Consolidation of Stable Long-Term Potentiation Lasting Months in the Hippocampus," *Journal of Neuroscience* 22, no. 21 (2002): 9626, https://doi.org/10.1523/JNEUROSCI.22-21-09626.2002.

119 **Exposure to a single dose of cocaine:** Mark A. Ungless, Jennifer L. Whistler, Robert C. Malenka, and Antonello Bonci, "Single Cocaine Exposure in Vivo Induces Long-Term Potentiation in Dopamine Neurons," *Nature* 411, no. 6837 (2001): 583–87, https://doi.org/10.1038/35079077.

119 **Stressing out a rat:** Lara M. Boyle, "A Neuroplasticity Hypothesis of Chronic Stress in the Basolateral Amygdala," *Yale Journal of Biology and Medicine* 86, no. 2 (2013): 117–25. https://pmc.ncbi.nlm.nih.gov/articles/PMC3670432.

119 **After an injury, the spinal cord:** Hidemasa Furue, Toshihiko Katafuchi, and Megumu Yoshimura, "Sensory Processing and Functional Reorganization of Sensory Transmission under Pathological Conditions in the Spinal Dorsal Horn," *Neuroscience Research* 48, no. 4 (2004): 361–68, https://doi.org/10.1016/j.neures.2003.12.005.

120 **Born in 1926, HM:** Larry R. Squire, "The Legacy of Patient H.M. for Neuroscience," *Neuron* 61, no. 1 (2009): 6–9, https://doi.org/10.1016/j.neuron.2008.12.023.

121 **Scoville drilled two holes:** W. B. Scoville, "Amnesia after Bilateral Mesial Temporal-Lobe Excision: Introduction to Case H.M," *Neuropsychologia* 6, no. 3 (1968): 211–13, https://doi.org/10.1016/0028-3932(68)90020-1.

121 **Archaeologists have unearthed:** Caroline Partiot, Aliénor Lepetit, Emilie Dodré, Camille Jenger, Bruno Maureille, Dominique Liguoro, and Aline Thomas, "Cranial Trepanation and Healing Process in Modern Patients—Bioarchaeological and Anthropological Implications," *Journal of Anatomy* 237, no. 6 (2020): 1049–61, https://doi.org/10.1111/joa.13266.

122 **slightly more than half:** Max O. Krucoff, Alvin Y. Chan, Stephen C. Harward, Shervin Rahimpour, John D. Rolston, Carrie Muh, and Dario J. Englot, "Rates and Predictors of Success and Failure in Repeat Epilepsy Surgery: A Meta-Analysis and Systematic Review," *Epilepsia* 58, no. 12 (2017): 2133–42, https://doi.org/10.1111/epi.13920.

122 **HM developed a severe case of amnesia:** W. B. Scoville and B. Milner, "Loss of Recent Memory after Bilateral Hippocampal Lesions. 1957," *Journal of Neuropsychiatry and Clinical Neurosciences* 12, no. 1 (2000): 103–13, https://doi.org/10.1176/jnp.12.1.103.

124 **Some types of memories require:** Suzanne Corkin, *Permanent Present Tense* (London: Penguin, 2013).

125 **A retired postal worker, Patient RB:** Nancy L. Rempel-Clower, Stuart M. Zola, Larry R. Squire, and David G. Amaral, "Three Cases of Enduring Memory Impairment after Bilateral Damage Limited to the Hippocampal Formation," *Journal of Neuroscience* 16, no. 16 (1996): 5233, https://doi.org/10.1523/JNEUROSCI.16-16-05233.1996.

125 **British musician Clive Wearing:** Oliver Sacks, "The Abyss," *The New Yorker*, September 17, 2007, https://www.newyorker.com/magazine/2007/09/24/the-abyss.

126 **Patient NA:** L. R. Squire, D. G. Amaral, S. Zola-Morgan, M. Kritchevsky, and G. Press, "Description of Brain Injury in the Amnesic Patient N.A. Based on Magnetic Resonance Imaging," *Experimental Neurology* 105, no. 1 (1989): 23–35, https://doi.org/10.1016/0014-4886(89)90168-4.

127 **Animals actively expend energy:** Pedro Bekinschtein, Noelia V. Weisstaub, Francisco Gallo, Maria Renner, and Michael C. Anderson, "A Retrieval-Specific Mechanism of Adaptive Forgetting in the Mammalian Brain," *Nature Communications* 9, no. 1 (2018): 4660, https://doi.org/10.1038/s41467-018-07128-7.

127 **food-gathering strategies never changed:** Briana Pobiner, "Evidence for Meat-Eating by Early Humans," *Nature Education Knowledge* 4, no. 6 (2013): 1, https://www.nature.com/scitable/knowledge/library/evidence-for-meat-eating-by-early-humans-103874273.

128 **risk for developing Korsakoff syndrome:** Nicolaas J. M. Arts, Serge J. W. Walvoort, and Roy P. C. Kessels, "Korsakoff's Syndrome: A Critical Review." *Neuropsychiatric Disease and Treatment* 13 (November 2017): 2875–90, https://doi.org/10.2147/NDT.S130078.

129 **Milder amnestic drugs:** Simon F. Crowe and Elizabeth K. Stranks, "The Residual Medium and Long-Term Cognitive Effects of Benzodiazepine Use: An Updated Meta-Analysis," *Archives of Clinical Neuropsychology* 33, no. 7 (2018): 901–11, https://doi.org/10.1093/arclin/acx120.

129 **Head trauma:** Christine N. Smith, Jennifer C. Frascino, Ramona O. Hopkins, and Larry R. Squire, "The Nature of Anterograde and Retrograde Memory Impairment after Damage to the Medial Temporal Lobe," *Neuropsychologia* 51, no. 13 (2013): 2709–14, https://doi.org/10.1016/j.neuropsychologia.2013.09.015.

130 **The landmark study was a 2005 paper:** R. Quian Quiroga, L. Reddy, G. Kreiman, C. Koch, and I. Fried, "Invariant Visual Representation by Single Neurons in the Human Brain." *Nature* 435, no. 7045 (2005): 1102–7, https://doi.org/10.1038/nature03687.

131 **by the ominous name KillerRed:** Timothy S. Jarvela and Adam D. Linstedt, "The Application of KillerRed for Acute Protein Inactivation in Living Cells," *Current Protocols in Cytometry* 69 (July 2014): 12.35.1–12.35.10, https://doi.org/10.1002/0471142956.cy1235s69.

133 **eyewitness accounts are sensitive to outside influences:** Moheb Costandi, "Evidence-Based Justice: Corrupted Memory," *Nature* 500, no. 7462 (2013): 268–70, https://doi.org/10.1038/500268a.

133 **"Three Truths and a Lie":** Elizabeth F. Loftus and Jacqueline E. Pickrell, "The Formation of False Memories," *Psychiatric Annals* 25, no. 12 (1995): 720–25, https://doi.org/10.3928/0048-5713-19951201-07.

134 **Nadean Cool:** Elizabeth F. Loftus, "Creating False Memories." *Scientific American* 277, no. 3 (1997): 70–75, http://www.jstor.org/stable/24995913.

135 **modeled after a laboratory experiment:** Kaja Głomb, "How to Improve Eyewitness Testimony Research: Theoretical and Methodological Concerns about Experiments on the Impact of Emotions on Memory Performance," *Psychological Research* 86, no. 1 (2022): 1–11, https://doi.org/10.1007/s00426-021-01488-4.

137 **incepted the memory of being afraid:** Xu Liu, Steve Ramirez, Petti T. Pang, Corey B. Puryear, Arvind Govindarajan, Karl Deisseroth, and Susumu Tonegawa, "Optogenetic Stimulation of a Hippocampal Engram Activates Fear Memory Recall," *Nature* 484, no. 7394 (2012): 381–85, https://doi.org/10.1038/nature11028.

137 **implanted a positive memory:** Steve Ramirez, Xu Liu, Christopher J. MacDonald, Anthony Moffa, Joanne Zhou, Roger L. Redondo, and Susumu Tonegawa, "Activating Positive Memory Engrams Suppresses Depression-Like Behaviour." *Nature* 522, no. 7556 (2015): 335–39, https://doi.org/10.1038/nature14514.

137 **Under the influence of a chemical:** Erwin Krediet, Tijmen Bostoen, Joost Breeksema, Annette van Schagen, Torsten Passie, and Eric Vermetten, "Reviewing the Potential of Psychedelics for the Treatment of PTSD," *International Journal of Neuropsychopharmacology* 23, no. 6 (2020): 385–400, https://doi.org/10.1093/ijnp/pyaa018.

CHAPTER 7

139 **Mark Twain:** Stanley Finger, "Mark Twain's Life-Long Fascination with Phrenology," *Journal of the History of the Behavioral Sciences* 55, no. 2 (2019): 99–121, https://doi.org/10.1002/jhbs.21960.

141 **Parts of Gall's writing were revolutionary:** Paul Eling and Stanley Finger, "Franz Joseph Gall on God and Religion: 'Dieu et Cerveau, Rien Que Dieu et Cerveau!,'"

Journal of the History of the Behavioral Sciences 58, no. 2 (2022): 183–203, https://doi.org/10.1002/jhbs.22151.

142 **Broca encountered the patient:** J. M. S. Pearce, "Broca's Aphasiacs," *European Neurology* 61, no. 3 (2009): 183–89, https://doi.org/10.1159/000189272.

144 **One hundred and forty years later:** N. F. Dronkers, O. Plaisant, M. T. Iba-Zizen, and E. A. Cabanis, "Paul Broca's Historic Cases: High Resolution MR Imaging of the Brains of Leborgne and Lelong," *Brain* 130, no. 5 (2007): 1432–41, https://doi.org/10.1093/brain/awm042.

146 **manual languages:** D. P. Corina, H. Poizner, U. Bellugi, T. Feinberg, D. Dowd, and L. O'Grady-Batch, "Dissociation between Linguistic and Nonlinguistic Gestural Systems: A Case for Compositionality," *Brain and Language* 43, no. 3 (1992): 414–47, https://doi.org/10.1016/0093-934x(92)90110-z.

146 **comprehension errors with Braille:** Annette K. Birchmeier, "Aphasic Dyslexia of Braille in a Congenitally Blind Man," *Neuropsychologia* 23, no. 2 (1985): 177–93, https://doi.org/10.1016/0028-3932(85)90102-2.

147 **situs inversus:** Herrera Ortiz, Andres Felipe, Juan C. Lacouture, Daniel Sandoval Medina, Luis J. Gómez Meléndez, and Rodolfo Uscategui, "Acute Cholecystitis in a Patient with Situs Inversus Totalis: An Unexpected Finding," *Cureus* 13, no. 6 (2021): e15799, https://doi.org/10.7759/cureus.15799.

147 **The Wada test involves injecting:** Syed Qadri, Hina Dave, Rohit Das, and Sasha Alick-Lindstrom, "Beyond the Wada: An Updated Approach to Pre-Surgical Language and Memory Testing: An Updated Review of Available Evaluation Techniques and Recommended Workflow to Limit Wada Test Use to Essential Clinical Cases," *Epilepsy Research* 174 (August 2021): 106673, https://doi.org/10.1016/j.eplepsyres.2021.106673.

148 **men typically scoring a higher:** Bennett A. Shaywitz, Sally E. Shaywitz, Kenneth R. Pugh, R. Todd Constable, Pawel Skudlarski, Robert K. Fulbright, Richard A. Bronen, Jack M. Fletcher, Donald P. Shankweiler, Leonard Katz, and John C. Gore, "Sex Differences in the Functional Organization of the Brain for Language," *Nature* 373, no. 6515 (1995): 607–9, https://doi.org/10.1038/373607a0.

148 **The biggest and most prominent:** J. Tomasch, "Size, Distribution, and Number of Fibres in the Human Corpus Callosum," *Anatomical Record* 119, no. 1 (1954): 119–35, https://doi.org/10.1002/ar.1091190109.

149 **they typically fare better in the long run:** Andrew Jea, Shobhan Vachhrajani, Elysa Widjaja, Daniel Nilsson, Charles Raybaud, Manohar Shroff, and James T. Rutka, "Corpus Callosotomy in Children and the Disconnection Syndromes: A Review," *Child's Nervous System* 24, no. 6 (2008): 685–92, https://doi.org/10.1007/s00381-008-0626-4.

150 **The most striking side effect:** Lisa A. Scepkowski and Alice Cronin-Golomb, "The Alien Hand: Cases, Categorizations, and Anatomical Correlates," *Behavioral and Cognitive Neuroscience Reviews* 2, no. 4 (2003): 261–77, https://doi.org/10.1177/1534582303260119.

151 **The sense of smell:** B. Eskenazi, W. S. Cain, E. D. Lipsitt, and R. A. Novelly, "Olfactory Functioning and Callosotomy: A Report of Two Cases," *Yale Journal*

of Biology and Medicine 61, no. 5 (1988): 447–56, https://pmc.ncbi.nlm.nih.gov/articles/PMC2590423.

152 **One patient, JW:** D. M. MacKay and Valerie MacKay, "Explicit Dialogue between Left and Right Half-Systems of Split Brains," *Nature* 295, no. 5851 (1982): 690–91, https://doi.org/10.1038/295690a0.

152 **She was using her chin:** Adrian Downey, "Split-Brain Syndrome and Extended Perceptual Consciousness," *Phenomenology and the Cognitive Sciences* 17, no. 4 (2018): 787–811, https://doi.org/10.1007/s11097-017-9550-y.

152 **In a separate "Name the Smell" experiment:** H. W. Gordon and R. W. Sperry, "Lateralization of Olfactory Perception in the Surgically Separated Hemispheres of Man," *Neuropsychologia* 7, no. 2 (1969): 111–20, https://doi.org/10.1016/0028-3932(69)90009-8.

153 **In the final paragraph:** R. W. Sperry, "Hemisphere Deconnection and Unity in Conscious Awareness," *American Psychologist* 23, no. 10 (1968): 723–33, https://doi.org/10.1037/h0026839.

153 **Some clues about the role of genetics in language:** K. E. Watkins, D. G. Gadian, and F. Vargha-Khadem, "Functional and Structural Brain Abnormalities Associated with a Genetic Disorder of Speech and Language," *American Journal of Human Genetics* 65, no. 5 (1999): 1215–21, https://doi.org/10.1086/302631.

154 **The gene provides the instructions:** Joery den Hoed, Karthikeyan Devaraju, and Simon E. Fisher, "Molecular Networks of the FOXP2 Transcription Factor in the Brain," *EMBO Reports* 22, no. 8 (2021): e52803, https://doi.org/10.15252/embr.202152803.

155 **Their songs use a vocabulary of specific patterns:** Samantha Carouso-Peck and Michael H. Goldstein, "Female Social Feedback Reveals Non-Imitative Mechanisms of Vocal Learning in Zebra Finches," *Current Biology* 29, no. 4 (2019): 631–636.e3, https://doi.org/10.1016/j.cub.2018.12.026.

155 **Artificially manipulating levels of FoxP2:** Lei Xiao, Devin P. Merullo, Therese M. I. Koch, Mou Cao, Marissa Co, Ashwinikumar Kulkarni, Genevieve Konopka, and Todd F. Roberts, "Expression of FoxP2 in the Basal Ganglia Regulates Vocal Motor Sequences in the Adult Songbird," *Nature Communications* 12, no. 1 (2021): 2617, https://doi.org/10.1038/s41467-021-22918-2.

155 **debate about the mammalian analogue:** Mark E. Hauber, Matthew I. M. Louder, and Simon C. Griffith, "Neurogenomic Insights into the Behavioral and Vocal Development of the Zebra Finch," *eLife* 10 (June 2021): e61849, https://doi.org/10.7554/eLife.61849.

156 **have harnessed this neuroinvasive property:** Jiamin Li, Taian Liu, Yun Dong, Kunio Kondoh, and Zhonghua Lu, "Trans-Synaptic Neural Circuit-Tracing with Neurotropic Viruses," *Neuroscience Bulletin* 35, no. 5 (2019): 909–20, https://doi.org/10.1007/s12264-019-00374-9.

156 **When symptoms do appear:** Charlotte James, Manale Harfouche, Nicky J. Welton, Katherine Me Turner, Laith J. Abu-Raddad, Sami L. Gottlieb, and Katharine J. Looker, "Herpes Simplex Virus: Global Infection Prevalence and Inci-

dence Estimates, 2016," *Bulletin of the World Health Organization* 98, no. 5 (2020): 315–29, https://doi.org/10.2471/BLT.19.237149.

157 **The pattern of injury is:** Ajith Kumar Ak, Beenish S. Bhutta, and Magda D. Mendez, "Herpes Simplex Encephalitis," in *StatPearls* (Treasure Island, FL: StatPearls Publishing, 2024), http://www.ncbi.nlm.nih.gov/books/NBK557643.

157 **probably came from Kansas:** John M. Barry, "The Site of Origin of the 1918 Influenza Pandemic and Its Public Health Implications," *Journal of Translational Medicine* 2, no. 1 (2004): 3, https://doi.org/10.1186/1479-5876-2-3.

158 **Von Economo had divided:** Lazaros C. Triarhou, "The Percipient Observations of Constantin von Economo on Encephalitis Lethargica and Sleep Disruption and Their Lasting Impact on Contemporary Sleep Research," *Brain Research Bulletin* 69, no. 3 (2006): 244–58, https://doi.org/10.1016/j.brainresbull.2006.02.002.

158 **Neurologist Dr. William Langston:** J. W. Langston and Jon Palfreman, *The Case of the Frozen Addicts: How the Solution of a Medical Mystery Revolutionized the Understanding of Parkinson's Disease* (Amsterdam: IOS Press, 2014).

159 **the designer drug MPPP:** Livia Pasquali, Elena Caldarazzo-Ienco, and Francesco Fornai, "MPTP Neurotoxicity: Actions, Mechanisms, and Animal Modeling of Parkinson's Disease," in *Handbook of Neurotoxicity*, ed. Richard M. Kostrzewa (New York: Springer, 2014), 237–75, https://doi.org/10.1007/978-1-4614-5836-4_170.

160 **On the upside, there was some good:** Susan H. Fox and Jonathan M. Brotchie, "Chapter 7—The MPTP-Lesioned Non-Human Primate Models of Parkinson's Disease. Past, Present, and Future," *Progress in Brain Research* 184 (2010): 133–57, https://doi.org/10.1016/S0079-6123(10)84007-5.

CHAPTER 8

161 **From paleolithic times:** Marc Azéma and Florent Rivère, "Animation in Palaeolithic Art: A Pre-Echo of Cinema," *Antiquity* 86, no. 332 (2012): 316–24, https://doi.org/10.1017/S0003598X00062785.

162 **detecting faces is highly adaptive:** Jiangang Liu, Jun Li, Lu Feng, Ling Li, Jie Tian, and Kang Lee, "Seeing Jesus in Toast: Neural and Behavioral Correlates of Face Pareidolia," *Cortex* 53 (April 2014): 60–77, https://doi.org/10.1016/j.cortex.2014.01.013.

162 **Sergent put participants in the PET scan:** Justine Sergent, Shinsuke Ohta, and Brennan Macdonald, "Functional Neuroanatomy of Face and Object Processing: A Positron Emission Tomography Study," *Brain* 115, no. 1 (1992): 15–36, https://doi.org/10.1093/brain/115.1.15.

164 **the FFA is activated more:** Gulsum Akdeniz, Sila Toker, and Ibrahim Atli, "Neural Mechanisms Underlying Visual Pareidolia Processing: An fMRI Study," *Pakistan Journal of Medical Sciences* 34, no. 6 (2018): 1560–66, https://doi.org/10.12669/pjms.346.16140.

164 **The harassment escalated:** Kenneth M. Heilman, "Justine Saade Sergent: Neuropsychologist Extraordinaire March 31, 1950–April 11, 1994," *Journal of the International Neuropsychological Society* 2, no. 5 (1996): 474, https://doi.org/10.1017/S1355617700001594.

164 **Renaissance painter Guiseppe Arcimboldo:** "Giuseppe Arcimboldo," 2018, https://www.theartstory.org/artist/arcimboldo-giuseppe.

166 **diagnostic tools to study:** Megumi Kobayashi, Yumiko Otsuka, Emi Nakato, So Kanazawa, Masami K. Yamaguchi, and Ryusuke Kakigi, "Do Infants Recognize the Arcimboldo Images as Faces? Behavioral and Near-Infrared Spectroscopic Study," *Journal of Experimental Child Psychology* 111, no. 1 (2012): 22–36, https://doi.org/10.1016/j.jecp.2011.07.008.

166 **a synesthete at a dinner party:** Richard E. Cytowic, *The Man Who Tasted Shapes*, rev. ed. (Cambridge, MA: MIT Press, 2008).

167 **there are more than a hundred different unions:** Richard E. Cytowic and Harry A. Whitaker, *Synesthesia* (New York: Springer, 1989), https://doi.org/10.1007/978-1-4612-3542-2.

167 **by environmental experiences and psycholinguistic rules:** Julia Simner, Jamie Ward, Monika Lanz, Ashok Jansari, Krist Noonan, Louise Glover, and David A. Oakley, "Non-Random Associations of Graphemes to Colours in Synaesthetic and Non-Synaesthetic Populations," *Cognitive Neuropsychology* 22, no. 8 (2005): 1069 85, https://doi.org/10.1080/02643290500200122.

168 **letter of the color name:** A. N. Rich, J. L. Bradshaw, and J. B. Mattingley, "A Systematic, Large-Scale Study of Synaesthesia: Implications for the Role of Early Experience in Lexical-Colour Associations," *Cognition* 98, no. 1 (2005): 53–84, https://doi.org/10.1016/j.cognition.2004.11.003.

168 **The first clinical description:** Jörg Jewanski, Sean A. Day, and Jamie Ward, "A Colorful Albino: The First Documented Case of Synaesthesia, by Georg Tobias Ludwig Sachs in 1812," *Journal of the History of the Neurosciences* 18, no. 3 (2009): 293–303, https://doi.org/10.1080/09647040802431946.

168 **Synesthesia is not considered pathological:** Duncan A. Carmichael, Rebecca Smees, Richard C. Shillcock, and Julia Simner, "Is There a Burden Attached to Synaesthesia? Health Screening of Synaesthetes in the General Population," *British Journal of Psychology* 110, no. 3 (2019): 530–48, https://doi.org/10.1111/bjop.12354.

169 **stronger connections between areas V4/V8:** Anna Zamm, Gottfried Schlaug, David M. Eagleman, and Psyche Loui, "Pathways to Seeing Music: Enhanced Structural Connectivity in Colored-Music Synesthesia," *NeuroImage* 74 (July 2013): 359–66, https://doi.org/10.1016/j.neuroimage.2013.02.024.

170 **Singer-songwriter Billie Eilish:** Josh Eells, "Billie Eilish and the Triumph of the Weird," *Rolling Stone*, July 31, 2019, https://www.rollingstone.com/music/music-features/billie-eilish-cover-story-triumph-weird-863603.

172 **Hemorrhagic stroke of the thalamus:** Tom A. Schweizer, Zeyu Li, Corinne E. Fischer, Michael P. Alexander, Stephen D. Smith, Simon J. Graham, and Luis Fornazarri, "From the Thalamus with Love: A Rare Window into the Locus of Emotional Synesthesia," *Neurology* 81, no. 5 (2013): 509–10, https://doi.org/10.1212/WNL.0b013e31829d86cc.

172 **Temporal lobe brain tumor:** Jean Vike, Bahman Jabbari, and Charles G. Maitland, "Auditory-Visual Synesthesia: Report of a Case with Intact Visual Path-

ways," *Archives of Neurology* 41, no. 6)1094): 680–81, https://doi.org/10.1001/archneur.1984.04210080092023.

172 **Amputation of an arm or leg:** Aviva I. Goller, Kerrie Richards, Steven Novak, and Jamie Ward, "Mirror-Touch Synaesthesia in the Phantom Limbs of Amputees," *Cortex* 49, no. 1 (2013): 243–51, https://doi.org/10.1016/j.cortex.2011.05.002.

173 **This effect is even seen among the Himba people:** Andrew J. Bremner, Serge Caparos, Jules Davidoff, Jan de Fockert, Karina J. Linnell, and Charles Spence, "'Bouba' and 'Kiki' in Namibia? A Remote Culture Make Similar Shape–Sound Matches, but Different Shape–Taste Matches to Westerners," *Cognition* 126, no. 2 (2013): 165–72, https://doi.org/10.1016/j.cognition.2012.09.007.

174 **Alice in Wonderland syndrome:** Jan Dirk Blom, "Alice in Wonderland Syndrome: A Systematic Review," *Neurology Clinical Practice* 6, no. 3 (2016): 259–70, https://doi.org/10.1212/CPJ.0000000000000251.

175 **decreases in blood flow:** Claudia Piervincenzi, Nikolaos Petsas, Alessandro Viganò, Valentina Mancini, Giulio Mastria, Marta Puma, Costanza Giannì, Vittorio Di Piero, and Patrizia Pantano, "Functional Connectivity Alterations in Migraineurs with Alice in Wonderland Syndrome," *Neurological Sciences* 44, no. 1 (2023): 305–17, https://doi.org/10.1007/s10072-022-06404-1.

175 **EEG activity:** Arya Shah, Setty M. Magaña, and Paul E. Youssef, "Do You See What I See? A Case of Alice in Wonderland Syndrome with EEG Correlate," *Child Neurology Open* 7 (January 2020): 2329048X20932714, https://doi.org/10.1177/2329048X20932714.

175 **prosopometamorphopsia is a visual illusory:** Sarah B. Herald, Jorge Almeida, and Brad Duchaine, "Face Distortions in Prosopometamorphopsia Provide New Insights into the Organization of Face Perception," *Neuropsychologia* 182 (April 2023): 108517, https://doi.org/10.1016/j.neuropsychologia.2023.108517.

176 **Melting zombie faces:** T. T. Winton-Brown, L. Smith, J. Laing, T. J. O'Brien, and A. Neal, "Ictal Face Perception Disturbance, Tattoos, and Reclaiming the Self," *Epilepsy and Behavior Reports* 22 (2023): 100595, https://doi.org/10.1016/j.ebr.2023.100595.

176 **Pablo Picasso's cubism period:** Joost Haan and Michel D. Ferrari, "Picasso's Migraine: Illusory Cubist Splitting or Illusion?," *Cephalalgia* 31, no. 9 (2011): 1057–60, https://doi.org/10.1177/0333102411406752.

176 **The skin of others' faces:** Jan Dirk Blom, Iris E. C. Sommer, Sanne Koops, and Oliver W. Sacks, "Prosopometamorphopsia and Facial Hallucinations," *Lancet* 384, no. 9958 (2014): 1998, https://doi.org/10.1016/S0140-6736(14)61690-1.

176 **Cases of prosopometamorphopsia:** Jan Dirk Blom, Bastiaan C. ter Meulen, Jitze Dool, and Dominic H. ffytche, "A Century of Prosopometamorphopsia Studies," *Cortex* 139 (June 2021): 298–308, https://doi.org/10.1016/j.cortex.2021.03.001.

177 **Also known as visual release hallucinations:** Thomas R. Hedges, "Charles Bonnet, His Life, and His Syndrome," *Survey of Ophthalmology* 52, no. 1 (2007): 111–14, https://doi.org/10.1016/j.survophthal.2006.10.007.

177 **20 percent of people:** Yousif Subhi, Diana Chabané Schmidt, Daniella Bach-Holm, Miriam Kolko, and Amardeep Singh, "Prevalence of Charles Bonnet

Syndrome in Patients with Glaucoma: A Systematic Review with Meta-Analyses," *Acta Ophthalmologica* 99, no. 2 (2021): 128–33, https://doi.org/10.1111/aos.14567.

178 **Lilliputian hallucinations:** Jan Dirk Blom, "Leroy's Elusive Little People: A Systematic Review on Lilliputian Hallucinations," *Neuroscience and Biobehavioral Reviews* 125 (June 2021): 627–36, https://doi.org/10.1016/j.neubiorev.2021.03.002.

178 **experiences under the influence of DMT:** Rick Strassman, "Subjective Effects of DMT and the Development of the Hallucinogen Rating Scale," 1992, https://maps.org/news-letters/v03n2/03208dmt.html.

179 **According to this thalamic filtering model:** Franz X. Vollenweider and Mark A. Geyer, "A Systems Model of Altered Consciousness: Integrating Natural and Drug-Induced Psychoses," *Brain Research Bulletin* 56, no. 5 (2001): 495–507, https://doi.org/10.1016/S0361-9230(01)00646-3.

179 **Just having the sensation:** Ben Alderson-Day, Peter Moseley, Kaja Mitrenga, Jamie Moffatt, Rebecca Lee, John Foxwell, Jacqueline Hayes, David Smailes, and Charles Fernyhough, "Varieties of Felt Presence? Three Surveys of Presence Phenomena and Their Relations to Psychopathology," *Psychological Medicine* 53, no. 8 (2023): 3692–700, https://doi.org/10.1017/S0033291722000344.

180 **feeling of an illusory companion:** Daniel Alejandro Drubach, "Twilight and Me: A Soliloquy," *Continuum* 27, no. 6 (2021): 1809–17, https://doi.org/10.1212/CON.0000000000001091.

180 **A hint comes from a study:** Shahar Arzy, Margitta Seeck, Stephanie Ortigue, Laurent Spinelli, and Olaf Blanke, "Induction of an Illusory Shadow Person," *Nature* 443, no. 7109 (2006): 287, https://doi.org/10.1038/443287a.

182 **One out of 100:** Natassia Robinson and Sarah E. Bergen, "Environmental Risk Factors for Schizophrenia and Bipolar Disorder and Their Relationship to Genetic Risk: Current Knowledge and Future Directions," *Frontiers in Genetics* 12 (June 2021): 686666, https://doi.org/10.3389/fgene.2021.686666.

183 **Persecutory delusions are:** Daniel Freeman and Philippa Garety, "Advances in Understanding and Treating Persecutory Delusions: A Review," *Social Psychiatry and Psychiatric Epidemiology* 49, no. 8 (2014): 1179–89, https://doi.org/10.1007/s00127-014-0928-7.

183 **Delusions of grandeur:** Louise Isham, Laura Griffith, Anne-Marie Boylan, Alice Hicks, Natalie Wilson, Rory Byrne, Bryony Sheaves, Richard P. Bentall, and Daniel Freeman, "Understanding, Treating, and Renaming Grandiose Delusions: A Qualitative Study," *Psychology and Psychotherapy* 94, no. 1 (2021): 119–40, https://doi.org/10.1111/papt.12260.

185 **A transcultural study:** T. M. Luhrmann, R. Padmavati, H. Tharoor, and A. Osei, "Differences in Voice-Hearing Experiences of People with Psychosis in the U.S.A., India and Ghana: Interview-Based Study," *British Journal of Psychiatry* 206, no. 1 (2015): 41–44, https://doi.org/10.1192/bjp.bp.113.139048.

185 **unofficial diagnosis of Truman Show syndrome:** Joel Gold and Ian Gold, "The 'Truman Show' Delusion: Psychosis in the Global Village," *Cognitive Neuropsychiatry* 17, no. 6 (2012): 455–72, https://pubmed.ncbi.nlm.nih.gov/22640240.

185 **A young Greek man:** Christos Mantas, Evgenia Papatheodorou, Maria-Elisavet Tsagkaropoulou, Anna Kourti, Georgios Georgiou, Petros Petrikis, and Thomas Hyphantis, "Delusions with Content Related to COVID-19 Pandemic, in Non-Infected Psychiatric Hospitalized Patients: A Six-Case Series," *Psychiatriki* 33, no. 4 (2022): 328–32, https://doi.org/10.22365/jpsych.2022.088.

185 **Another man developed persecutory delusions:** Esra Aydin Sunbul, Emine Cengiz Cavusoglu, and Huseyin Gulec, "Brief Psychotic Disorder during COVID-19 Pandemic: A Case Series," *Indian Journal of Psychiatry* 63, no. 5 (2021): 508–10, https://doi.org/10.4103/indianjpsychiatry.indianjpsychiatry_1130_20.

186 **Interpersonal family dynamics:** A. K. Kala and N. N. Wig, "Delusion across Cultures," *International Journal of Social Psychiatry* 28, no. 3 (1982): 185–93, https://doi.org/10.1177/002076408202800304.

186 **prompted by loneliness:** Ralph E. Hoffman, "A Social Deafferentation Hypothesis for Induction of Active Schizophrenia," *Schizophrenia Bulletin* 33, no. 5 (2007): 1066–70, https://doi.org/10.1093/schbul/sbm079.

186 **A woman avoided showers:** Hans Debruyne, Michael Portzky, Frédérique Van den Eynde, and Kurt Audenaert, "Cotard's Syndrome: A Review," *Current Psychiatry Reports* 11, no. 3 (2009): 197–202, https://doi.org/10.1007/s11920-009-0031-z.

186 **A middle-aged woman discontinued:** Anne Ruminjo and Boris Mekinulov, "A Case Report of Cotard's Syndrome," *Psychiatry* 5, no. 6 (2008): 28–29, https://pmc.ncbi.nlm.nih.gov/articles/PMC2695744.

187 **a 1788 manuscript:** Hans Förstl and Barbara Beats, "Charles Bonnet's Description of Cotard's Delusion and Reduplicative Paramnesia in an Elderly Patient (1788)," *British Journal of Psychiatry* 160, no. 3 (1992): 416–18, https://doi.org/10.1192/bjp.160.3.416.

187 **the condition was given its official name:** G. E. Berrios and R. Luque, "Cotard's Delusion or Syndrome? A Conceptual History," *Comprehensive Psychiatry* 36, no. 3 (1995): 218–23, https://doi.org/10.1016/0010-440x(95)90085-a.

CHAPTER 9

189 **A metal rod:** Malcolm Macmillan and Phineas Gage, *An Odd Kind of Fame: Stories of Phineas Gage* (Cambridge, MA: MIT Press, 2022).

190 **when Gage forcefully rammed:** Ricardo Vieira Teles, "Phineas Gage's Great Legacy," *Dementia and Neuropsychologia* 14, no. 4 (2020): 419–21, https://doi.org/10.1590/1980-57642020dn14-040013.

190 **Gage was not as shocked:** J. M. Harlow and Edgar Miller, "Recovery from the Passage of an Iron Bar through the Head," *History of Psychiatry* 4, no. 14 (1993): 271–73, https://doi.org/10.1177/0957154X9300401406.

190 **Less than a month after his accident:** Alan G. Lewandowski, Joshua D. Weirick, Caroline A. Lewandowski, and Jack Spector, "1079Phineas Gage: A Neuropsychological Perspective of a Historical Case Study," in *The Oxford Handbook of the History of Clinical Neuropsychology*, edited by William B. Barr and Linas A. Bieliauskas (Oxford: Oxford University Press, 2024). https://doi.org/10.1093/oxfordhb/9780199765683.013.21.

192 **modern imaging techniques:** Hanna Damasio, Thomas Grabowski, Randall Frank, Albert M. Galaburda, and Antonio R. Damasio, "The Return of Phineas Gage: Clues about the Brain from the Skull of a Famous Patient," *Science* 264, no. 5162 (1994): 1102–5, https://doi.org/10.1126/science.8178168.

193 **formation of a brain tumor:** Florien W. Boele, Alasdair G. Rooney, Robin Grant, and Martin Klein, "Psychiatric Symptoms in Glioma Patients: From Diagnosis to Management," *Neuropsychiatric Disease and Treatment* 11 (March 2015): 1413–20, https://doi.org/10.2147/NDT.S65874.

193 **recorded evidence of new and unusual volatility:** Maurice Heatly, "Text of Psychiatrist's Notes on Sniper," archive.nytimes.com/www.nytimes.com/library/national/080366tx-shoot.html?TB_iframe=true&height=921.6&width=921.6.

194 **excerpts from Whitman's suicide notebooks:** Cara Santa Maria, "WATCH: The Mind of a Mass Murderer," March 8, 2012, https://www.huffpost.com/entry/mind-murderer_n_1384102.

194 **one pediatric patient:** Intractable Seizures, Compulsions, and Coprolalia: A Pediatric Case Study," *Journal of Neuropsychiatry and Clinical Neurosciences* 4, no. 3 (1992): 315–19, https://psychiatryonline.org/doi/abs/10.1176/jnp.4.3.315.

194 **according to one estimation:** Sreenivasa R. Chandana, Sujana Movva, Madan Arora, and Trevor Singh, "Primary Brain Tumors in Adults," *American Family Physician* 77, no. 10 (2008): 1423–30, https://www.aafp.org/pubs/afp/issues/2008/0515/p1423.pdf.

195 **1991 case of Herbert Weinstein:** Kevin Davis, *The Brain Defense: Murder in Manhattan and the Dawn of Neuroscience in America's Courtrooms* (New York: Penguin, 2017).

195 **in chronic traumatic encephalopathy:** Philip H. Montenigro, Daniel T. Corp, Thor D. Stein, Robert C. Cantu, and Robert A. Stern, "Chronic Traumatic Encephalopathy: Historical Origins and Current Perspective," *Annual Review of Clinical Psychology* 11, no. 1 (2015): 309–30, https://doi.org/10.1146/annurev-clinpsy-032814-112814.

195 **NFL player Aaron Hernandez?:** Alexandra L. Aaronson, Sean D. Bordelon, S. Jan Brakel, and Helen Morrison, "A Review of the Role of Chronic Traumatic Encephalopathy in Criminal Court," *Journal of the American Academy of Psychiatry and the Law Online* 49, no. 1 (2021): 60, https://doi.org/10.29158/JAAPL.200054-20.

195 **consider John Hinckley Jr.'s trial:** Lincoln Caplan, *The Insanity Defense and the Trial of John W. Hinckley, Jr.* (Boston: D. R. Godine, 1984).

197 **hand dominance:** Polly Henninger, "Conditional Handedness: Handedness Changes in Multiple Personality Disordered Subject Reflect Shift in Hemispheric Dominance," *Consciousness and Cognition* 1, no. 3 (1992): 265–87, https://doi.org/10.1016/1053-8100(92)90065-I.

197 **or the ability to see:** Hans Strasburger and Bruno Waldvogel, "Sight and Blindness in the Same Person: Gating in the Visual System," *PsyCh Journal* 4, no. 4 (2015): 178–85, https://doi.org/10.1002/pchj.109.

197 **heavy cannabis use predict:** Emma C. Johnson et al., "A Large-Scale Genome-Wide Association Study Meta-Analysis of Cannabis Use Disorder," *Lancet Psychiatry* 7, no. 12 (2020): 1032–45, https://doi.org/10.1016/S2215-0366(20)30339-4.

197 **2009 financial crisis in Greece:** Stelios Stylianidis and Kyriakos Souliotis, "The Impact of the Long-Lasting Socioeconomic Crisis in Greece," *BJPsych International* 16, no. 1 (2019): 16–18, https://doi.org/10.1192/bji.2017.31.

198 **the patient experiences severe or prolonged abuse:** Constance J. Dalenberg, Bethany L. Brand, David H. Gleaves, Martin J. Dorahy, Richard J. Loewenstein, Etzel Cardeña, Paul A. Frewen, Eve B. Carlson, and David Spiegel, "Evaluation of the Evidence for the Trauma and Fantasy Models of Dissociation," *Psychological Bulletin* 138, no. 3 (2012): 550–88, https://doi.org/10.1037/a0027447.

198 **socio-cognitive model:** Debbie Nathan, *Sybil Exposed: The Extraordinary Story behind the Famous Multiple Personality Case* (New York: Free Press, 2012).

199 **According to Cornelia Wilbur:** Mark Miller, "Unmasking Sybil," *Newsweek*, January 24, 1999, https://www.newsweek.com/unmasking-sybil-165174.

199 **At the root of the controversy:** Ben Harris, "Sybil, Inc.," *Science* 334, no. 6054 (2011): 312, https://doi.org/10.1126/science.1212843.

199 **Anatomically, these patients have:** David Blihar, Elliott Delgado, Marina Buryak, Michael Gonzalez, and Randall Waechter, "A Systematic Review of the Neuroanatomy of Dissociative Identity Disorder," *European Journal of Trauma and Dissociation* 4, no. 3 (2020): 100148, https://doi.org/10.1016/j.ejtd.2020.100148.

199 **also seen in PTSD and major depressive disorder:** Mark Nolan, Elena Roman, Anurag Nasa, Kirk J. Levins, Erik O'Hanlon, Veronica O'Keane, and Darren Willian Roddy, "Hippocampal and Amygdalar Volume Changes in Major Depressive Disorder: A Targeted Review and Focus on Stress," *Chronic Stress* 4 (September 2020): 2470547020944553, https://doi.org/10.1177/2470547020944553.

200 **one model of psychopathology:** Lauren A. N. Lebois, Poornima Kumar, Cori A. Palermo, Ashley M. Lambros, Lauren O'Connor, Jonathan D. Wolff, Justin T. Baker, Staci A. Gruber, Nina Lewis-Schroeder, Kerry J. Ressler, Matthew A. Robinson, Sherry Winternitz, Lisa D. Nickerson, and Milissa L. Kaufman, "Deconstructing Dissociation: A Triple Network Model of Trauma-Related Dissociation and Its Subtypes," *Neuropsychopharmacology* 47, no. 13 (2022): 2261–70, https://doi.org/10.1038/s41386-022-01468-1.

200 **Billy Milligan was on trial:** Daniel Keyes, *The Minds of Billy Milligan* (New York: Random House, 1981).

200 **DID defense didn't work for Kenneth Bianchi:** Philip M. Coons, "Iatrogenesis and Malingering of Multiple Personality Disorder in the Forensic Evaluation of Homicide Defendants," *Multiple Personality Disorder* 14, no. 3 (1991): 757–68, https://doi.org/10.1016/S0193-953X(18)30299-5.

202 **According to her most famous theory:** H. E. Fisher, "Lust, Attraction, and Attachment in Mammalian Reproduction," *Human Nature* 9, no. 1 (1998): 23–52, https://doi.org/10.1007/s12110-998-1010-5.

203 **only 9 percent of mammalian species:** D. Lukas and T. H. Clutton-Brock, "The Evolution of Social Monogamy in Mammals," *Science* 341, no. 6145 (2013): 526–30, https://doi.org/10.1126/science.1238677.

203 **Once a pair of wild voles mate:** Lisa A. McGraw and Larry J. Young, "The Prairie Vole: An Emerging Model Organism for Understanding the Social

Brain," *Trends in Neurosciences* 33, no. 2 (2010): 103–9, https://doi.org/10.1016/j.tins.2009.11.006.

203 **In the partner preference test:** Kyle Gobrogge and Zuoxin Wang, "The Ties That Bond: Neurochemistry of Attachment in Voles," *Current Opinion in Neurobiology* 38, (June 2016): 80–88, https://doi.org/10.1016/j.conb.2016.04.011.

204 **Male montane voles mate:** T. R. Insel and L. E. Shapiro, "Oxytocin Receptor Distribution Reflects Social Organization in Monogamous and Polygamous Voles," *Proceedings of the National Academy of Sciences of the United States of America* 89, no. 13 (1992): 5981–85, https://doi.org/10.1073/pnas.89.13.5981.

205 **misattribution of arousal:** D. G. Dutton and A. P. Aron, "Some Evidence for Heightened Sexual Attraction under Conditions of High Anxiety," *Journal of Personality and Social Psychology* 30, no. 4 (1974): 510–17, https://doi.org/10.1037/h0037031.

208 **Enmark called out to the gunman:** Christopher Klein, "The Birth of "Stockholm Syndrome: The True Story of Hostages Loyal to Their Captor," August 23, 2018, https://www.history.com/news/stockholm-syndrome.

208 **difficult time leaving abusive relationships:** Rebecca Bailey, Jaycee Dugard, Stefanie F. Smith, and Stephen W. Porges, "Appeasement: Replacing Stockholm Syndrome as a Definition of a Survival Strategy," *European Journal of Psychotraumatology* 14, no. 1 (2023): 2161038, https://doi.org/10.1080/20008066.2022.2161038.

209 **labor pain amnesia:** Catherine A. Niven and Tricia Murphy-Black, "Memory for Labor Pain: A Review of the Literature," *Birth* 27, no. 4 (2000): 244–53, https://doi.org/10.1046/j.1523-536x.2000.00244.x.

209 **one of the more dangerous activities:** Ai-Ris Y. Collier and Rose L. Molina, "Maternal Mortality in the United States: Updates on Trends, Causes, and Solutions," *NeoReviews* 20, no. 10 (2019): e561–74, https://doi.org/10.1542/neo.20-10-e561.

210 **hearing the cries:** Silvana Valtcheva, Habon A. Issa, Chloe J. Bair-Marshall, Kathleen A. Martin, Kanghoon Jung, Yiyao Zhang, Hyung-Bae Kwon, and Robert C. Froemke, "Neural Circuitry for Maternal Oxytocin Release Induced by Infant Cries," *Nature* 621, no. 7980 (2023): 788–95, https://doi.org/10.1038/s41586-023-06540-4.

210 **baby cries caused a huge increase in amygdala activity:** Madelon M. E. Riem, Marian J. Bakermans-Kranenburg, Suzanne Pieper, Mattie Tops, Maarten A. S. Boksem, Robert R. J. M. Vermeiren, Marinus H. van IJzendoorn, and Serge A. R. B. Rombouts, "Oxytocin Modulates Amygdala, Insula, and Inferior Frontal Gyrus Responses to Infant Crying: A Randomized Controlled Trial," *Genotype, Circuits, and Cognition in Autism and Attention-Deficit/Hyperactivity Disorder* 70, no. 3 (2011): 291–97, https://doi.org/10.1016/j.biopsych.2011.02.006.

211 **extends even into early childhood:** Kalina J. Michalska, Jean Decety, Chunyu Liu, Qi Chen, Meghan E. Martz, Suma Jacob, Alison E. Hipwell, Steve S. Lee, Andrea Chronis-Tuscano, and Irwin D. Waldman, "Genetic Imaging of the Association of Oxytocin Receptor Gene (OXTR) Polymorphisms with Positive Maternal Parenting," *Frontiers in Behavioral Neuroscience* 8 (February 2014), https://doi.org/10.3389/fnbeh.2014.00021.

211 **gender-dependent differences in specific tasks:** Sally A. Grace, Susan L. Rossell, Markus Heinrichs, Catarina Kordsachia, and Izelle Labuschagne, "Oxytocin and Brain Activity in Humans: A Systematic Review and Coordinate-Based Meta-Analysis of Functional MRI Studies," *Psychoneuroendocrinology* 96 (October 2018): 6–24, https://doi.org/10.1016/j.psyneuen.2018.05.031.

CHAPTER 10

214 **A psychiatrist even published a case report:** J. C. Bozzuto, "Cinematic Neurosis Following 'The Exorcist': Report of Four Cases," *Journal of Nervous and Mental Disease* 161, no. 1 (1975): 43–48, https://doi.org/10.1097/00005053-197507000-00005.

215 **For the Kaluli people:** Deborah A. Elliston, "Erotic Anthropology: 'Ritualized Homosexuality' in Melanesia and Beyond," *American Ethnologist* 22, no. 4 (1995): 848–67, http://www.jstor.org/stable/646389.

215 **After the Port Said riot:** "Egypt Football: Death Sentences over Port Said Stadium Violence," January 26, 2013, https://www.bbc.com/news/world-middle-east-21209808.

216 **Today, totalitarian:** Timo Kivimäki, "When Ideologies Became Dangerous: An Analysis of the Transformation of the Relationship between Security and Oppositional Ideologies in US Presidential Discourse," *Global Society* 37, no. 2 (2022): 225–44, https://www.tandfonline.com/doi/full/10.1080/13600826.2022.2061923.

216 **Researchers recruited indigenous Dutch:** Carsten K. W. De Dreu, Lindred L. Greer, Gerben A. Van Kleef, Shaul Shalvi, and Michel J. J. Handgraaf, "Oxytocin Promotes Human Ethnocentrism," *Proceedings of the National Academy of Sciences of the United States of America* 108, no. 4 (2011): 1262–66, https://doi.org/10.1073/pnas.1015316108.

217 **the ORE was first described:** Gustave A. Feingold, "The Influence of Environment on Identification of Persons and Things," *Journal of the American Institute of Criminal Law and Criminology* 5, no. 1 (1914): 39, https://doi.org/10.2307/1133283.

218 **A little snort of oxytocin:** Iris Blandón-Gitlin, Kathy Pezdek, Sesar Saldivar, and Erin Steelman, "Oxytocin Eliminates the Own-Race Bias in Face Recognition Memory," *Brain Research* 1580 (September 2014): 180–87, https://doi.org/10.1016/j.brainres.2013.07.015.

218 **taboo in practically every culture:** D. T. Max, *The Family That Couldn't Sleep: A Medical Mystery* (New York: Random House, 2007).

219 **200 people die each year:** R. W. Hornabrook and D. J. Moir, "KURU," *Lancet* 296, no. 7684 (1970): 1175–79, https://doi.org/10.1016/S0140-6736(70)90356-9.

221 **The Fore believed that everyone:** Jerome T. Whitfield, Wandagi H. Pako, John Collinge, and Michael P. Alpers, "Mortuary Rites of the South Fore and Kuru," *Philosophical Transactions of the Royal Society of London. Series B, Biological Sciences* 363, no. 1510 (2008): 3721–24, https://doi.org/10.1098/rstb.2008.0074.

221 **They also detail his pedophilia:** Mike McCarthy, "Nobel Prize Winner Gajdusek Admits Child Abuse," *Lancet* 349, no. 9052 (1997): 623, https://doi.org/10.1016/S0140-6736(97)23009-6.

221 **during the outbreak of mad cow disease:** Dennis Normile, "First U.S. Case of Mad Cow Sharpens Debate over Testing," *Science* 303, no. 5655 (2004): 156–57, https://doi.org/10.1126/science.303.5655.156.

222 **Millions of cattle were slaughtered:** Thierry Billette de Villemeur, "Chapter 124—Creutzfeldt–Jakob Disease," *Handbook of Clinical Neurology* 112 (2013): 1191–93, https://doi.org/10.1016/B978-0-444-52910-7.00040-4.

222 **Carefully separating out the brain and spinal cord:** Diane L. Ritchie, Alexander H. Peden, and Marcelo A. Barria, "Variant CJD: Reflections a Quarter of a Century On," *Pathogens* 10, no. 11 (2021): 1413, https://doi.org/10.3390/pathogens10111413.

222 ***kuru* free:** Michael P. Alpers, "A History of Kuru," *Papua and New Guinea Medical Journal* 50, nos. 1–2 (2007): 10–19, https://pubmed.ncbi.nlm.nih.gov/19354007.

224 **more genetic variation:** Theresa M. Duello, Shawna Rivedal, Colton Wickland, and Annika Weller, "Race and Genetics versus 'Race' in Genetics: A Systematic Review of the Use of African Ancestry in Genetic Studies," *Evolution, Medicine, and Public Health* 9, no. 1 (2021): 232–45, https://doi.org/10.1093/emph/eoab018.

224 **justified the annexation:** Elise Juzda, "Skulls, Science, and the Spoils of War: Craniological Studies at the United States Army Medical Museum, 1868–1900," *Studies in History and Philosophy of Science Part C: Studies in History and Philosophy of Biological and Biomedical Sciences* 40, no. 3 (2009): 156–67, https://doi.org/10.1016/j.shpsc.2009.06.010.

224 **researchers developed the Implicit Association Test:** Anthony G. Greenwald, Miguel Brendl, Huajian Cai, Dario Cvencek, John F. Dovidio, Malte Friese, Adam Hahn, Eric Hehman, Wilhelm Hofmann, Sean Hughes, Ian Hussey, Christian Jordan, Teri A. Kirby, Calvin K. Lai, Jonas W. B. Lang, Kristen P. Lindgren, Dominika Maison, Brian D. Ostafin, James R. Rae, Kate A. Ratliff, Adriaan Spruyt, and Reinout W. Wiers, "Best Research Practices for Using the Implicit Association Test," *Behavior Research Methods* 54, no. 3 (2022): 1161–80, https://doi.org/10.3758/s13428-021-01624-3.

226 **hemodynamic changes pop up:** Kristine M. Knutson, Linda Mah, Charlotte F. Manly, and Jordan Grafman, "Neural Correlates of Automatic Beliefs about Gender and Race," *Human Brain Mapping* 28, no. 10 (2007): 915–30, https://doi.org/10.1002/hbm.20320.

227 **which informs us about body ownership:** Jakub Limanowski, Antoine Lutti, and Felix Blankenburg, "The Extrastriate Body Area Is Involved in Illusory Limb Ownership," *NeuroImage* 86 (February 2014): 514–24, https://doi.org/10.1016/j.neuroimage.2013.10.035.

227 **seen in psychopaths:** Kent A. Kiehl, "A Cognitive Neuroscience Perspective on Psychopathy: Evidence for Paralimbic System Dysfunction," *Psychiatry Research* 142, nos. 2–3 (2006): 107–28, https://doi.org/10.1016/j.psychres.2005.09.013.

227 **In the same-race condition:** Ruben T. Azevedo, Emiliano Macaluso, Alessio Avenanti, Valerio Santangelo, Valentina Cazzato, and Salvatore Maria Aglioti, "Their Pain Is Not Our Pain: Brain and Autonomic Correlates of Empathic Resonance with the Pain of Same and Different Race Individuals," *Human Brain Mapping* 34, no. 12 (2013): 3168–81, https://doi.org/10.1002/hbm.22133.

228 **pervasiveness of these falsehoods:** Kelly M. Hoffman, Sophie Trawalter, Jordan R. Axt, and M. Norman Oliver, "Racial Bias in Pain Assessment and Treatment Recommendations, and False Beliefs about Biological Differences between Blacks and Whites," *Proceedings of the National Academy of Sciences of the United States of America* 113, no. 16 (2016): 4296–4301, https://doi.org/10.1073/pnas.1516047113.

228 **undertreated with analgesics:** Randall W. Knoebel, Janet V. Starck, and Pringl Miller, "Treatment Disparities among the Black Population and Their Influence on the Equitable Management of Chronic Pain," *Health Equity* 5, no. 1 (2021): 596–605, https://doi.org/10.1089/heq.2020.0062.

229 **Belonging to a minoritized population:** Nathalie M. Dumornay, Lauren A. M. Lebois, Kerry J. Ressler, and Nathaniel G. Harnett, "Racial Disparities in Adversity during Childhood and the False Appearance of Race-Related Differences in Brain Structure," *American Journal of Psychiatry* 180, no. 2 (2023): 127–38, doi:10.1176/appi.ajp.21090961.

229 **Some minoritized populations:** Sarah K. Letang, Shayne S.-H. Lin, Patricia A. Parmelee, and Ian M. McDonough, "Ethnoracial Disparities in Cognition Are Associated with Multiple Socioeconomic Status-Stress Pathways," *Cognitive Research: Principles and Implications* 6, no. 1 (2021): 64, https://doi.org/10.1186/s41235-021-00329-7.

230 **but the disease is rare:** Oye Gureje, Adesola Ogunniyi, Olusegun Baiyewu, Brandon Price, Frederick W. Unverzagt, Rebecca M. Evans, Valerie Smith-Gamble, Kathleen A. Lane, Sujuan Gao, Kathleen S. Hall, Hugh C. Hendrie, and Jill R. Murrell, "*APOE* ε4 Is Not Associated with Alzheimer's Disease in Elderly Nigerians," *Annals of Neurology* 59, no. 1 (2006): 182–85, https://doi.org/10.1002/ana.20694.

230 **Living in a community:** Rachel A. Rabin and Lena Palaniyappan, "Brain Health in Ethnically Minority Youth at Risk for Psychosis," *Neuropsychopharmacology* 48, no. 12 (2023): 1701–2, https://doi.org/10.1038/s41386-023-01719-9.

230 **combat racial disparities in research:** Tricia Choy, Elizabeth Baker, and Katherine Stavropoulos, "Systemic Racism in EEG Research: Considerations and Potential Solutions," *Affective Science* 3, no. 1 (2022): 14–20, https://doi.org/10.1007/s42761-021-00050-0.

230 **is skin complexion:** Jasmine Kwasa, Hannah M. Peterson, Kavon Karrobi, Lietsel Jones, Termara Parker, Nia Nickerson, and Sossena Wood, "Demographic Reporting and Phenotypic Exclusion in fNIRS," *Frontiers in Neuroscience* 17 (May 2023): 1086208, https://doi.org/10.3389/fnins.2023.1086208.

231 **The UK Biobank contains:** Cathie Sudlow, John Gallacher, Naomi Allen, Valerie Beral, Paul Burton, John Danesh, Paul Downey, Paul Elliott, Jane Green, Martin Landray, Bette Liu, Paul Matthews, Giok Ong, Jill Pell, Alan Silman, Alan Young, Tim Sprosen, Tim Peakman, and Rory Collins, "UK Biobank: An Open Access Resource for Identifying the Causes of a Wide Range of Complex Diseases of Middle and Old Age," *PLoS Medicine* 12, no. 3 (2015): e1001779, https://doi.org/10.1371/journal.pmed.1001779.

232 **Of the psychology studies published:** Joseph Henrich, Steven J. Heine, and Ara Norenzayan, "The Weirdest People in the World?," *Behavioral and Brain*

Sciences 33, nos. 2–3 (2010): 61–83; discussion 83–135, https://doi.org/10.1017/S0140525X0999152X.

233 **But there is one behavior:** Benedikt Herrmann, Christian Thöni, and Simon Gächter, "Antisocial Punishment across Societies," *Science* 319, no. 5868 (2008): 1362–67, https://doi.org/10.1126/science.1153808.

233 **judged groupings of abstract shapes:** Ara Norenzayan, Edward E. Smith, Beom Jun Kim, and Richard E. Nisbett, "Cultural Preferences for Formal versus Intuitive Reasoning," *Cognitive Science* 26, no. 5 (2002): 653–84, https://doi.org/10.1207/s15516709cog2605_4.

233 **The Müller-Lyer effect:** Anton Killin and Ross Pain, "How WEIRD Is Cognitive Archaeology? Engaging with the Challenge of Cultural Variation and Sample Diversity," *Review of Philosophy and Psychology* 14, no. 2 (2023): 539–63, https://doi.org/10.1007/s13164-021-00611-z.

234 **heavily industrialized infrastructures:** R. L. Gregory, "Distortion of Visual Space as Inappropriate Constancy Scaling," *Nature* 199, no. 4894 (1963): 678–80, https://doi.org/10.1038/199678a0.

236 **contact hypothesis:** Thomas F. Pettigrew and Linda R. Tropp, "A Meta-Analytic Test of Intergroup Contact Theory," *Journal of Personality and Social Psychology* 90, no. 5 (2006): 751–83, https://doi.org/10.1037/0022-3514.90.5.751.

236 **U.S. Army during World War II:** John F. Dovidio, Samuel L. Gaertner, and Kerry Kawakami, "Intergroup Contact: The Past, Present, and the Future," *Group Processes and Intergroup Relations* 6, no. 1 (2003): 5–21, https://doi.org/10.1177/1368430203006001009.

236 **product more of acculturation:** S. Sangrigoli, C. Pallier, A.-M. Argenti, V. A. G. Ventureyra, and S. de Schonen, "Reversibility of the Other-Race Effect in Face Recognition during Childhood," *Psychological Science* 16, no. 6 (2005): 440–44, https://doi.org/10.1111/j.0956-7976.2005.01554.x.

237 **clever rat model for empathy:** Inbal Ben-Ami Bartal, David A. Rodgers, Maria Sol Bernardez Sarria, Jean Decety, and Peggy Mason, "Pro-Social Behavior in Rats Is Modulated by Social Experience," *eLife* 3 (January 2014): e01385, https://doi.org/10.7554/eLife.01385.

238 **Pain is also emotionally contagious in mice:** Dale J. Langford, Sara E. Crager, Zarrar Shehzad, Shad B. Smith, Susana G. Sotocinal, Jeremy S. Levenstadt, Mona Lisa Chanda, Daniel J. Levitin, and Jeffrey S. Mogil, "Social Modulation of Pain as Evidence for Empathy in Mice," *Science* 312, no. 5782 (2006): 1967–70, https://doi.org/10.1126/science.1128322.

238 **consider weaver ants:** F. M. K. Uy, J. D. Adcock, S. F. Jeffries, and E. Pepere, "Intercolony Distance Predicts the Decision to Rescue or Attack Conspecifics in Weaver Ants," *Insectes Sociaux* 66, no. 2 (2019): 185–92, https://doi.org/10.1007/s00040-018-0674-z.

CHAPTER 11

240 **free will discourse in 1983:** B. Libet, C. A. Gleason, E. W. Wright, and D. K. Pearl, "Time of Conscious Intention to Act in Relation to Onset of Cerebral Activity

(Readiness-Potential): The Unconscious Initiation of a Freely Voluntary Act," *Brain* 106, no. 3 (1983): 623–42, https://doi.org/10.1093/brain/106.3.623.

242 **terrifying reproductive behavior:** V. Velev, M. Dinkova, and A. Mirtschew, "Human Botfly Infection," *QJM* 112, no. 7 (2019): 529–30, https://doi.org/10.1093/qjmed/hcz089.

243 **99 percent of spider species:** Aristos Georgiou, "These Are the World's Most Dangerous Spiders," August 31, 2022, https://www.newsweek.com/worlds-most-dangerous-spiders-1738569.

243 **wasp finds an orb-weaver spider:** William G. Eberhard, "The Natural History and Behavior of *Hymenoepimecis argyraphaga* (Hymenoptera: Ichneumonidae) a Parasitoid of *Plesiometa argyra* (Araneae: Tetragnathidae)," *Journal of Hymenoptera Research* 9, no. 2 (2000): 220–40.

245 **Once inside the snails:** Vishvas Gowda, Susha Dinesh, and Sameer Sharma, "Manipulative Neuroparasites: Uncovering the Intricacies of Neurological Host Control," *Archives of Microbiology* 205, no. 9 (2023): 314, https://doi.org/10.1007/s00203-023-03637-2.

245 **practically every avian and mammalian species:** Audrey Simon, Michel Bigras Poulin, Alain N. Rousseau, and Nicholas H. Ogden, "Fate and Transport of *Toxoplasma gondii* Oocysts in Seasonally Snow Covered Watersheds: A Conceptual Framework from a Melting Snowpack to the Canadian Arctic Coasts," *International Journal of Environmental Research and Public Health* 10, no. 3 (2013): 994–1005, https://doi.org/10.3390/ijerph10030994.

245 **when they play outside:** Scott R. Loss, Tom Will, and Peter P. Marra, "The Impact of Free-Ranging Domestic Cats on Wildlife of the United States," *Nature Communications* 4, no. 1 (2013): 1396, https://doi.org/10.1038/ncomms2380.

246 **one side of the box contains a dish of bobcat urine:** Wendy Marie Ingram, Leeanne M. Goodrich, Ellen A. Robey, and Michael B. Eisen, "Mice Infected with Low-Virulence Strains of *Toxoplasma gondii* Lose Their Innate Aversion to Cat Urine, Even after Extensive Parasite Clearance," *PLoS One* 8, no. 9 (2013): e75246, https://doi.org/10.1371/journal.pone.0075246.

246 **Mice become less fearful:** Shiraz Tyebji, Simona Seizova, Anthony J. Hannan, and Christopher J. Tonkin, "Toxoplasmosis: A Pathway to Neuropsychiatric Disorders," *Neuroscience and Biobehavioral Reviews* 96 (January 2019): 72–92, https://doi.org/10.1016/j.neubiorev.2018.11.012.

246 **in the basolateral amygdala:** Rupshi Mitra, Robert Morris Sapolsky, and Ajai Vyas, "*Toxoplasma gondii* Infection Induces Dendritic Retraction in Basolateral Amygdala Accompanied by Reduced Corticosterone Secretion," *Disease Models and Mechanisms* 6, no. 2 (2013): 516–20, https://doi.org/10.1242/dmm.009928.

247 **we pick it up from drinking contaminated water:** J. G. Montoya and O. Liesenfeld, "Toxoplasmosis," *Lancet* 363, no. 9425 (2004): 1965–76, https://doi.org/10.1016/S0140-6736(04)16412-X.

247 **Antipsychotic medications such as haloperidol:** Guillaume Fond, Alexandra Macgregor, Ryad Tamouza, Nora Hamdani, Alexandre Meary, Marion Leboyer, and Jean-Francois Dubremetz, "Comparative Analysis of Anti-Toxoplasmic

Activity of Antipsychotic Drugs and Valproate," *European Archives of Psychiatry and Clinical Neuroscience* 264, no. 2 (2014): 179–83, https://doi.org/10.1007/s00406-013-0413-4.

247 **parasite is a risk factor in the disease:** Despina G. Contopoulos-Ioannidis, Maria Gianniki, Angeline Ai-Nhi Truong, and Jose G. Montoya, "Toxoplasmosis and Schizophrenia: A Systematic Review and Meta-Analysis of Prevalence and Associations and Future Directions," *Psychiatric Research and Clinical Practice* 4, no. 2 (2022): 48–60, https://doi.org/10.1176/appi.prcp.20210041.

248 **The stereotype needs to be updated:** Jitka Lindová, Martina Novotná, Jan Havlíček, Eva Jozífková, Anna Skallová, Petra Kolbeková, Zdeněk Hodný, Petr Kodym, and Jaroslav Flegr, "Gender Differences in Behavioural Changes Induced by Latent Toxoplasmosis," *International Journal for Parasitology* 36, no. 14 (2006): 1485–92, https://doi.org/10.1016/j.ijpara.2006.07.008.

248 **psychomotor consequences:** J. Havlíček, Z. Gašová, A. P. Smith, K. Zvára, and J. Flegr, "Decrease of Psychomotor Performance in Subjects with Latent 'Asymptomatic' Toxoplasmosis," *Parasitology* 122, no. 5 (2001): 515–20, https://doi.org/10.1017/S0031182001007624.

248 **impulsivity and aggression and decreases rule following:** Thomas B. Cook, Lisa A. Brenner, C. Robert Cloninger, Patricia Langenberg, Ajirioghene Igbide, Ina Giegling, Annette M. Hartmann, Bettina Konte, Marion Friedl, Lena Brundin, Maureen W. Groer, Adem Can, Dan Rujescu, and Teodor T. Postolache, "'Latent' Infection with *Toxoplasma gondii*: Association with Trait Aggression and Impulsivity in Healthy Adults," *Journal of Psychiatric Research* 60 (January 2015): 87–94, https://doi.org/10.1016/j.jpsychires.2014.09.019.

248 **everyday task that requires good reaction time:** Shaban Gohardehi, Mehdi Sharif, Shahabeddin Sarvi, Mahmood Moosazadeh, Reza Alizadeh-Navaei, Seyed Abdollah Hosseini, Afsaneh Amouei, Abdolsattar Pagheh, Mitra Sadeghi, and Ahmad Daryani, "The Potential Risk of Toxoplasmosis for Traffic Accidents: A Systematic Review and Meta-Analysis," *Experimental Parasitology* 191 (August 2018): 19–24, https://doi.org/10.1016/j.exppara.2018.06.003.

251 **Pascuala mural in Spain:** Brian P. Akers, Juan Francisco Ruiz, Alan Piper, and Carl A. P. Ruck, "A Prehistoric Mural in Spain Depicting Neurotropic Psilocybe Mushrooms?," *Economic Botany* 65, no. 2 (2011): 121–28, https://doi.org/10.1007/s12231-011-9152-5.

251 **in pre-Columbian Mesoamerican societies:** F. J. Carod-Artal, "Hallucinogenic Drugs in Pre-Columbian Mesoamerican Cultures," *Neurología (English Edition)* 30, no. 1 (2015): 42–49, https://doi.org/10.1016/j.nrleng.2011.07.010.

252 **Project MK Ultra:** *Project MKUltra, the CIA's Program of Research in Behavioral Modification*, Joint Hearing before the Select Committee on Intelligence and the Subcommittee on Health and Scientific Research of the Committee on Human Resources, United States Senate, 95th Congress, First Session (Washington, DC: Government Printing Office, 1977).

252 **by giving LSD:** Stephen Kinzer, *Poisoner in Chief: Sidney Gottlieb and the CIA Search for Mind Control* (New York: Henry Holt, 2019).

253 **Researchers discovered that the ChR2:** Georg Nagel, Tanjef Szellas, Wolfram Huhn, Suneel Kateriya, Nona Adeishvili, Peter Berthold, Doris Ollig, Peter Hegemann, and Ernst Bamberg, "Channelrhodopsin-2, a Directly Light-Gated Cation-Selective Membrane Channel," *Proceedings of the National Academy of Sciences* 100, no. 24 (2004): 13940–45, https://doi.org/10.1073/pnas.1936192100.

253 **milliseconds after exposure to light:** Edward S. Boyden, Feng Zhang, Ernst Bamberg, Georg Nagel, and Karl Deisseroth, "Millisecond-Timescale, Genetically Targeted Optical Control of Neural Activity," *Nature Neuroscience* 8, no. 9 (2005): 1263–68, https://doi.org/10.1038/nn1525.

254 **they move quickly toward the food:** Yexica Aponte, Deniz Atasoy, and Scott M. Sternson, "AGRP Neurons Are Sufficient to Orchestrate Feeding Behavior Rapidly and without Training," *Nature Neuroscience* 14, no. 3 (2011): 351–55, https://doi.org/10.1038/nn.2739.

254 **A two-room box:** Zisis Bimpisidis, Niclas König, and Åsa Wallén-Mackenzie, "Two Different Real-Time Place Preference Paradigms Using Optogenetics within the Ventral Tegmental Area of the Mouse," *Journal of Visualized Experiments*, no. 156 (February 2020), https://doi.org/10.3791/60867.

254 **He suddenly jolts awake:** Roberto De Luca, Stefano Nardone, Kevin P. Grace, Anne Venner, Michela Cristofolini, Sathyajit S. Bandaru, Lauren T. Sohn, Dong Kong, Takatoshi Mochizuki, Bianca Viberti, Lin Zhu, Antonino Zito, Thomas E. Scammell, Clifford B. Saper, Bradford B. Lowell, Patrick M. Fuller, and Elda Arrigoni, "Orexin Neurons Inhibit Sleep to Promote Arousal," *Nature Communications* 13, no. 1 (2022): 4163, https://doi.org/10.1038/s41467-022-31591-y.

255 **a significant alleviation of this abnormal pain:** Robert P. Bonin, Feng Wang, Mireille Desrochers-Couture, Alicja Gąsecka, Marie-Eve Boulanger, Daniel C. Côté, and Yves De Koninck, "Epidural Optogenetics for Controlled Analgesia," *Molecular Pain* 12 (January 2016): 1744806916629051, https://doi.org/10.1177/1744806916629051.

255 **The bird changes their song:** Lei Xiao, Devin P. Merullo, Therese M. I. Koch, Mou Cao, Marissa Co, Ashwinikumar Kulkarni, Genevieve Konopka, and Todd F. Roberts, "Expression of FoxP2 in the Basal Ganglia Regulates Vocal Motor Sequences in the Adult Songbird," *Nature Communications* 12, no. 1 (2021): 2617, https://doi.org/10.1038/s41467-021-22918-2.

255 **the two-arm bandit task:** Johannes Passecker, Nace Mikus, Hugo Malagon-Vina, Philip Anner, Jordane Dimidschstein, Gordon Fishell, Georg Dorffner, and Thomas Klausberger, "Activity of Prefrontal Neurons Predict Future Choices during Gambling," *Neuron* 101, no. 1 (2019): 152–164.e7, https://doi.org/10.1016/j.neuron.2018.10.050.

257 **The drug phenobarbital:** Cassaundra B. Lewis, Preeti Patel, and Ninos Adams, "Phenobarbital," in *StatPearls* (Treasure Island, FL: StatPearls Publishing, 2024), http://www.ncbi.nlm.nih.gov/books/NBK532277.

258 **mapping patient experiences:** Wilder Penfield and Edwin Boldrey, "Somatic Motor and Sensory Representation in the Cerebral Cortex of Man as Studied by Electrical Stimulation," *Brain* 60, no. 4 (1937): 389–443, https://doi.org/10.1093/brain/60.4.389.

259 **Stimulating other parts of the cortex:** Kevin A. Mazurek and Marc H. Schieber, "How Is Electrical Stimulation of the Brain Experienced, and How Can We Tell? Selected Considerations on Sensorimotor Function and Speech," *Cognitive Neuropsychology* 36, nos. 3–4 (2019): 103–16, https://doi.org/10.1080/02643294.2019.1609918.

259 **Robert Heath's main:** Lone Frank, *The Pleasure Shock: The Rise of Deep Brain Stimulation and Its Forgotten Inventor* (New York: Dutton, 2018).

259 **legacy was tarnished:** Christen M. O'Neal, Cordell M. Baker, Chad A. Glenn, Andrew K. Conner, and Michael E. Sughrue, "Dr. Robert G. Heath: A Controversial Figure in the History of Deep Brain Stimulation," *Neurosurgical Focus* 43, no. 3 (2017): E12, https://doi.org/10.3171/2017.6.FOCUS17252.

260 **Patient B-19:** R. G. Heath, "Pleasure and Brain Activity in Man: Deep and Surface Electroencephalograms during Orgasm," *Journal of Nervous and Mental Disease* 154, no. 1 (1972): 3–18, https://doi.org/10.1097/00005053-197201000-00002.

261 **brain stimulation tools have been approved:** Samantha L. Cohen, Marom Bikson, Bashar W. Badran, and Mark S. George, "A Visual and Narrative Timeline of US FDA Milestones for Transcranial Magnetic Stimulation (TMS) Devices," *Brain Stimulation* 15, no. 1 (2022): 73–75, https://doi.org/10.1016/j.brs.2021.11.010.

261 **In preclinical studies:** Jared C. Horvath, Jennifer M. Perez, Lachlan Forrow, Felipe Fregni, and Alvaro Pascual-Leone, "Transcranial Magnetic Stimulation: A Historical Evaluation and Future Prognosis of Therapeutically Relevant Ethical Concerns," *Journal of Medical Ethics* 37, no. 3 (2011): 137, https://doi.org/10.1136/jme.2010.039966.

262 **The left inferior frontal gyrus:** Tali Sharot, Ryota Kanai, David Marston, Christoph W. Korn, Geraint Rees, and Raymond J. Dolan, "Selectively Altering Belief Formation in the Human Brain," *Proceedings of the National Academy of Sciences* 109, no. 42 (2012): 17058–62, https://doi.org/10.1073/pnas.1205828109.

263 **neurostimulation of the hippocampus:** Elva Arulchelvan and Sven Vanneste, "Promising Neurostimulation Routes for Targeting the Hippocampus to Improve Episodic Memory: A Review," *Brain Research* 1815 (September 2023): 148457, https://doi.org/10.1016/j.brainres.2023.148457.

263 **participants answered quicker:** Roberta Sellaro, Belle Derks, Michael A. Nitsche, Bernhard Hommel, Wery P. M. Van Den Wildenberg, Kristina Van Dam, and Lorenza S. Colzato, "Reducing Prejudice through Brain Stimulation," *Brain Stimulation* 8, no. 5 (2015): 891–97, https://doi.org/10.1016/j.brs.2015.04.003.

CLOSING COMMENTS

265 **disorder that causes uncontrolled movements:** Lazzaro di Biase, Pasquale Maria Pecoraro, Simona Paola Carbone, Maria Letizia Caminiti, and Vincenzo Di Lazzaro, "Levodopa-Induced Dyskinesias in Parkinson's Disease: An Overview on Pathophysiology, Clinical Manifestations, Therapy Management Strategies and Future Directions," *Journal of Clinical Medicine* 12, no. 13 (2023): 4427, https://doi.org/10.3390/jcm12134427.

268 **advocacy training:** Elizabeth D. Hetherington and Alexandra A. Phillips, "A Scientist's Guide for Engaging in Policy in the United States," *Frontiers in Marine Science* 7 (June 2020): 409, https://doi.org/10.3389/fmars.2020.00409.

268 **he established the Village Landais:** Marion Pech, Céline Meillon, Manon Marquet, Jean-François Dartigues, and Hélène Amieva, "The 'Alzheimer Village': Assessment of Alzheimer's Disease Representations in the General Population: A Cross Sectional Phone Survey," *Alzheimer's and Dementia* 8, no. 1 (2022): e12328, https://doi.org/10.1002/trc2.12328.

270 **even identical twins:** Tinca J. C. Polderman, Beben Benyamin, Christiaan A. de Leeuw, Patrick F. Sullivan, Arjen van Bochoven, Peter M. Visscher, and Danielle Posthuma, "Meta-Analysis of the Heritability of Human Traits Based on Fifty Years of Twin Studies," *Nature Genetics* 47, no. 7 (2015): 702–9, https://doi.org/10.1038/ng.3285.

INDEX